European Radiation Protection Course

Basics

Philippe Massiot and Christine Jimonet
Coordinators

Cover illustrations: L. Godart/CEA

Printed in France
ISBN: 978-2-7598-0703-1

This work is subject to copyright. All rights are reserved, whether the whole or part of the material is concerned, specifically the rights of translation, reprinting, re-use of illustrations, recitation, broadcasting, reproduction on microfilms or in other ways, and storage in data bank. Duplication of this publication or parts thereof is only permitted under the provisions of the French Copyright law of March 11, 1957. Violations fall under the prosecution act of the French Copyright law.
© EDP Sciences 2014

Preface: European Network on Education and Training in RAdiological Protection II-ENETRAP II

Radiation protection is a major challenge when using ionising radiation, both in nuclear and non-nuclear industries, as well as in other areas such as healthcare and research. Therefore, maintaining a high level of competence in radiation protection is crucial to ensure the protection of man and environment and to ensure the development of new technologies in a safe way.

Within the European 7FP project ENETRAP II, specific attention is given to the development of radiation protection training, with the view to maximising transfer of high-level knowledge and understanding. As in all 7FP projects in the area of education and training in nuclear fission, safety and radiation protection, emphasis is put on multi-disciplinary and transnational and inter-sectorial mobility. The ultimate goal is to contribute to a European system for continuous professional development, which relies on the principles of modularity of courses and common qualification criteria, a common mutual recognition system, and facilitating lecturer, learner and worker mobility across the EU.

This text book is developed in the frame of ENETRAP II and supports radiation protection training for Radiation Protection Experts (RPEs), and for any other person dealing with ionising radiation in their daily practice.

The topics treated are in line with the requirements for RPEs as stated in the new EURATOM Basic Safety Standards. They reflect the content of the generic modules of the European Reference Training Scheme for RPEs that forms an essential basis for the implementation of mutual recognition of RPEs through Europe.

This book contains the theoretical background of radiation protection principles and invites the learner to implement the acquired knowledge in daily work situations via exercises. In addition, QR code is added that guides the learner to supplementary on-line exercises. An e-book complements this text book and provides continuously updated exercises and simulations of practical situations for which the RPE must be able to advise on the radiation protection measures to be taken.

We wish you interesting reading.

Michèle Coeck
ENETRAP II Coordinator
On behalf of the ENETRAP II Consortium

Foreword: Textbook, cyberbook and ECVET

The European ENETRAP II (European Network on Education and Training in Radiological Protection II) project was made up of several parts, one of which was focused on the development of a textbook. After an analysis of nearly 60 books on radiation protection, it was decided to write a textbook combining both theory and exercises where the reader becomes responsible for his own learning.

In this text book, you will find the first chapters of Module 1 of the training of Radiation Protection Expert (RPE) where the definition and missions are defined in the new European Directive. At the end of each chapter, exercises allow you to assess yourself.

In addition, the QR code sends you to a site where additional resources such as exercises and corrections enable you to develop the concepts outlined in the textbook.

Cyberbook QR code

This educational resource, sometimes called an e-book, aims to offer the reader additional resources. This site is based on Moodle (Learning Management System) LMS widely used by project partners.

During the project, we were asked to implement the ECVET (European Credit for Vocational Education and Training) approach. This approach aims to promote mobility within Europe through a process of recognition of acquired skills and mutual confidence.

Each competence is characterized by the following three descriptors: knowledge, skills and attitudes.

Thus, the RPE training is described in the e-book by about 80 skills and about 400 learning outcomes. Training is therefore driven by the expected skills and not by the content of training provided. So, the question becomes implicit "what is acquired and not what are the subjects taught".

We hope you enjoy reading this book that aims to be in some way a precursor of a series to come.

Paul Livolsi
Head of the WP7 and WP4 of ENETRAP II

Authors

Marc AMMERICH Advanced Technician in Radiation Protection, CNAM engineer in nuclear physics and holds a Master in Aerosols Science. After beginning his career in the Department of Radiological Protection at the CEA, Saclay, he joined the radiation protection team of the INSTN in 1991 and became its manager in 1996. Working at the French Nuclear Safety Authority (ASN) in 2001, he is a confirmed inspector in radiation protection. After having joined the direction of Protection and Nuclear Safety of the CEA in 2006, he has served as a nuclear inspector at the CEA since 2008. He is also a lecturer at the INSTN and communicating-researcher. He received the SFEN award in 1989 for the realization of ICARE bench.

Jean-Christophe BODINEAU holds a BTS in Radiation Protection, he began his career at the CEA by monitoring the X-ray generators, particle accelerators and radiation sources used in research and industry. This experience has allowed him to teach the practice of radiation protection. He became an engineer in nuclear science and technology and he began to teach radioactivity, interaction and radiation detection at the INSTN. After obtaining a Master in Applied Physics, he became responsible for teaching in the institute. He specializes in teaching the detection of ionising radiation, a field in which he is recognized as a senior expert at the CEA. An important part of his activities for the training of doctors and nurses of the French nuclear power plants in the field of anthropogammametry.

Hugues BRUCHET obtained his master's degree in 2001 (DESS Radioprotection, University of Grenoble). Currently he is an engineer at the CEA (French Atomic Energy and alternative energies Commission) and deputy head of the teaching unit "Health Technologies and Radiation Protection" at the INSTN (National Institute for Nuclear Science and Technology part of the CEA). Moreover, he is involved in teaching the "Personnes compétentes en radioprotection (PCR)" as a certified teacher, author and coordinator of books in the field of radiation protection, and member of PCR teachers certification comitee (CEFRI).

Cécile ETARD	Medical physicist. After 2 years in a radiotherapy department, Cecile Etard joined the Central Laboratory for Electrical Industries where she practiced for six years as an engineer in the metrology of ionising radiation. She joined the INSTN in 2000 as head teacher and trainer of several lessons in the field of radiation protection and medical physics. In 2003, she undertook the responsibility of the INSTN Radiation Protection team. In 2007, she joined the Unit of Expertise in Medical Radiation Institute for Radiological Protection and Nuclear Safety (IRSN).
Christine JIMONET	PhD in Biochemistry, graduated from the University of Paris XI, she is in charge of the unit "Technology for Health and Radiation Protection" within the National Institute for Nuclear Sciences and Techniques (INSTN) at the CEA. More specifically she taught the topic "Biological effects of ionising radiations" in various training courses. At the INSTN, she is also the manager of education related to the medical internship in Nuclear Medicine.
Philippe MASSIOT	CNAM Engineer in Nuclear Science and Technology, Philippe Massiot began as a researcher at the CEA of particle accelerators in the field of materials and radiobiology. He then specialized in radio toxicology of actinides. After 5 years spent in the French Nuclear Safety Authority (ASN) as training manager and radiation protection inspector, he is now responsible for teaching Radiation Protection and involved in a European project to harmonize regulatory training in radiation protection. He is also an expert at the CEA.
Henri MÉTIVIER	PhD, is a former Research Director at the CEA. Former member of the ICRP, Professor Emeritus at the INSTN, Chairman of "Comité de rédaction de la revue Radioprotection", Journal of the French Society for Radiation Protection, Chairman of the Drafting Committee for Radiation Protection. He is the author and coordinator of numerous books in the field of radiation protection and also plutonium.
Jean-Claude MOREAU	Radiation Protection Technician, Engineer CNAM in Physics, he worked in radiation at CEA/Saclay for 16 years and held leadership positions at STMI (Areva group). After a brief detour in environmental technology, in 2000 he founded the company CAP2i, a firm specializing in radiation protection studies, expertise and training. He has taught radiation at the INSTN, in several universities and trained many "Personnes compétentes en radioprotection (PCR)".

Abdel-Mijd NOURREDINE — PhD in Physical Sciences and Professor at the University Louis Pasteur of Strasbourg I. He operates a multidisciplinary research institute Hubert-Curien (IPHC), where he leads the leadership team Radiation and Environmental Measures (RaMsEs). Specialist in subatomic physics and nuclear applications, he has a rich experience in Education and PCR training. He has supervised several PhD thesis on R&D in nuclear instrumentation and dosimetry of ionising radiation.

Hervé VIGUIER — CNAM Engineer in Nuclear Science and Technology, Hervé Viguier is a research engineer at the French Atomic Energy Commission and alternative energies. He is a training officer and trainer in the field of radioactivity, radiation protection and detection of ionising radiation. He is also in charge of different practical work of various engineering courses.

Alain VIVIER — Engineering School of the Air and engineer in nuclear engineering with weapons option. After several years operating in the unit of nuclear weapon systems on the Plateau d'Albion, he taught physics and nuclear measurement at the School of Military Applications of Atomic Energy (EAMEA). He joined the CEA as head of the radiation protection group assigned to the Plutonium team in Cadarache. He then joined the INSTN Saclay where he set up a training session on ionising radiation dosimetry, among others.

Acknowledgments

This book has benefited in various ways (proofreading, iconography) from the contribution to the appointees below for which they are sincerely thanked.

Thomas Berkvens	SCK•CEN Academy, Education and Training, Belgium
Jean-Marc Bordy	CEA, DRT/LIST/DM2I/LNHB, France
Marjan Moreels	SCK•CEN, Radiation Department Biology, Belgium
Francois Paquet	IRSN, Direction de la Stratégie, du Développement et des Partenariats, France
Maria-Joao Santiago-Ribeiro	CHRU Tours, France
Jennifer Tavassoli	The English Network, France

Contents

Preface: European Network on Education and Training
in RAdiological Protection II-ENETRAP II i

Foreword: Textbook, cyberbook and ECVET iii

Authors v

Acknowledgments viii

Chapitre 1. Radioactivity and nuclear physics

- 1.1. General considerations 2
  - 1.1.1. The structure of matter 2
  - 1.1.2. Definitions. Nomenclature 4
  - 1.1.3. Isotopes and isobars 5
- 1.2. Nuclear stability and instability 5
  - 1.2.1. Stable nuclei 6
  - 1.2.2. Radioactive nuclei 6
- 1.3. Radiation: energy and emission intensity 7
  - 1.3.1. Radiation energy 7
  - 1.3.2. Radiation: emission intensity 7
- 1.4. Nuclear transformation modes 8
  - 1.4.1. Radioactive disintegration modes 8
  - 1.4.2. Gamma decay 16
  - 1.4.3. Metastable nuclei 17
- 1.5. The electron cloud 18
  - 1.5.1. Configuration of an electron cloud 18
  - 1.5.2. Spontaneous rearrangement 22
  - 1.5.3. Induced rearrangementt 24
- 1.6. Decay schemes 24
- 1.7. Physical quantities and fundamental properties 25

	1.7.1.	Activity	25
	1.7.2.	Emission rate	26
	1.7.3.	Radioactive decay and half-life	26
	1.7.4.	Radioactive series	28
	1.7.5.	Activity–mass relationship	32
	1.7.6.	Production of radionuclides	33
1.8.	Check your knowledge		36

Chapitre 2. Interaction of ionising radiation with matter

2.1.	Ionizing radiation: definition and classification		42
2.2.	Interaction of charged particles with matter		44
	2.2.1.	General considerations	44
	2.2.2.	Interaction of electrons with matter	45
	2.2.3.	Interaction of heavy charged particles with matter: the case of alpha particles	53
2.3.	Interaction of electromagnetic radiation with matter		55
	2.3.1.	The photoelectric effect	56
	2.3.2.	The Compton effect	57
	2.3.3.	Pair production	59
	2.3.4.	Attenuation of electromagnetic radiation	60
2.4.	Interaction of neutrons with matter		66
	2.4.1.	General considerations	66
	2.4.2.	Neutron absorption	67
	2.4.3.	Neutron scattering	69
	2.4.4.	Neutron attenuation	69
2.5.	Check your knowledge		71

Chapitre 3. Dosimetry

3.1.	Physical quantities		77
	3.1.1.	Absorbed dose	78
	3.1.2.	Relation between dose and fluence	79
	3.1.3.	Dose calculations for charged particles	79
	3.1.4.	Dose calculations for γ- and X-photons	86
3.2.	Protection quantities		93
	3.2.1.	Equivalent dose	93
	3.2.2.	Effective dose	94
3.3.	Operational quantities		94
3.4.	Check your knowledge		98

Chapitre 4. Biological effects of ionising radiation

4.1.	Molecular effects of interaction with ionising radiation	106
4.2.	Cellular effects, consequences of molecular effects	110

4.3.	Non-stochastic or deterministic effects		112
	4.3.1.	Effects of localised irradiation	113
	4.3.2.	Effects of a single, global and homogeneous irradiation of the entire organism	115
	4.3.3.	Characteristics of deterministic effects	117
4.4.	Stochastic effects		118
4.5.	Summary		119
4.6.	Risk assessment		120
	4.6.1.	Carcinogenic effects	120
	4.6.2.	Genetic effects	122
	4.6.3.	Quantification of the total risk of stochastic effects	122
	4.6.4.	The concept of radiation detriment	122
4.7.	The principles of the ICRP		126
4.8.	Check your Knowledge		128

Chapitre 5. Detection and measurement of ionising radiation

5.1.	Detectors		132
	5.1.1.	Scintillation counters	132
	5.1.2.	Gas-filled detectors	137
	5.1.3.	Semiconductor detectors	141
	5.1.4.	Photographic emulsions	144
	5.1.5.	Radioluminescent detectors	145
	5.1.6.	Other types of detectors	148
5.2.	Electronics associated with detectors		150
5.3.	Measurement methods and practices		153
	5.3.1.	Detection pulse counting	153
	5.3.2.	Measurement of an ionisation current	180
	5.3.3.	Integrating ionisations over the duration of exposure: passive detectors	182
5.4.	Check your Knowledge		187

Chapitre 6. Uses of sources of ionising radiation

6.1.	Natural sources of ionising radiation		191
	6.1.1.	Cosmic radiation	191
	6.1.2.	Telluric radiation	192
6.2.	Medical applications of ionising radiation		195
	6.2.1.	Diagnosis	200
	6.2.2.	Therapy	200
	6.2.3.	Other equipment	200
6.3.	Industrial applications of ionising radiation		201
	6.3.1.	Industrial radiography	201
	6.3.2.	Metrology and analysis devices	203
	6.3.3.	Industrial irradiators	208

	6.3.4.	Miscellaneous uses of radionuclides as sealed sources.............	209

6.3.4. Miscellaneous uses of radionuclides as sealed sources 209
6.3.5. Uses of radionuclides as unsealed sources in industry and research 209
6.4. Civil nuclear industry 210
6.4.1. Nuclear fuel 210
6.4.2. Uranium ore extraction 211
6.4.3. Nuclear fuel fabrication 211
6.4.4. "Pressurised Water Reactor" type nuclear reactor 213
6.4.5. Nuclear fuel reprocessing 215

Bibliography 217

Radioactivity and nuclear physics

Hugues Bruchet, Marc Ammerich, Cécile Etard, Hervé Viguier, Abdel-Mjid Nourreddine

Introduction

Radioactivity is the property, exhibited by some nuclei, of transforming into one or more new nuclei, while emitting – in that transformation – a helium nucleus (i.e. an alpha particle), an electron (beta particle), or electromagnetic radiation (gamma radiation).

Radioactivity is a natural phenomenon, which was discovered, at the close of the 19th century, by French physicist Henri Becquerel. Investigating the phenomenon of phosphorescence, he sought to find out whether the radiation emitted by phosphorescent uranium salts was to be identified with the X-rays discovered by German physicist Wilhelm Roentgen, the preceding year. He showed that a photographic plate could become clouded through the agency of such salts, without first exposing these to any light. He came to the conclusion, therefore, that uranium spontaneously emits radiation that has the ability to cloud a photographic plate, quite apart from any phosphorescence process.

To refer to this phenomenon, Pierre and Marie Curie coined the term "radioactivity." In the months that followed the discovery Henri Becquerel had made, Marie Curie showed that, in like manner to uranium, thorium is naturally radioactive. Subsequently, working with several tonnes of uranium oxide ore, the Curies were able to isolate first polonium, then radium – a chemical element that is 2.5 million times more highly radioactive than uranium.

Radioactivity is an integral part of atomic physics, this being the science concerned with the study of the phenomena inherent in the atomic nucleus, and its constituents. Consequently, the present chapter begins with a review, describing the basic constituents of matter, and setting out the nomenclature in use. Thereafter, the phenomenon of radioactive decay, and the associated processes are described, and detailed. Finally, definitions are given for the fundamental physical quantities and properties involved, in particular the activity of a radioactive source, its half-life, and the concept of a radioactive decay series, to round out the chapter.

1.1. General considerations

1.1.1. *The structure of matter*

In nature, matter – whether it be air, water, stars, living organisms… – consists of molecules, which in turn are combinations of atoms. As early as classical Antiquity, Greek philosophers averred that matter is made up of minute "building blocks," combining with one another. The present-day word "atom" indeed has come down from that time, being derived from the Greek *atomos,* meaning "that which cannot be cut, indivisible."

> Things retain their substance unimpaired, till a powerful enough
> force be found to come upon them, in proportion to their structure.
> Nothing whatever, therefore, recedes into nothingness, but all things,
> being rent asunder, turn back into the elements of matter. …
> Nothing at all, then, is seen to pass away utterly,
> since Nature recruits one thing from another.
>
> Lucretius (99–55 BCE), *De rerum natura (On the Nature of Things),* Bk. 1, 246–263

In an atom, two components may be distinguished: the nucleus, at the center, and the electron cloud.

- the central nucleus consists of an assembly of two kinds of particle: **protons**, and **neutrons**, also known, collectively, as nucleons;
- the electron cloud consists of an ensemble of **electrons**, orbiting the nucleus at high speed. Mathematical formulae are the only means allowing the regions to be determined, where electrons are most likely to be found, in the cloud they form around the nucleus. Such regions are known as "electron shells;" despite the uncertainty inherent in any electron's position, the localization of these regions is nonetheless fairly precise, and this so-called "shell" model – while altogether inaccurate by present standards – does make it possible to account, fairly simply, for the physical phenomena that arise.

The atom's electron cloud is spherical, with a diameter of the order of 10^{-10} meter. The nucleus is smaller still, since it fills a sphere some 10^{-14} meter in diameter, on average – in other words, it is 10 000 times smaller than the sphere containing the atom as a whole. The huge gap extending between the nucleus and the electrons is empty: taking an atom's nucleus to be as a football placed in the middle of a sports ground, then the electrons would be seen as tiny marbles around the stands.

The atom's mass is not distributed evenly across the atom. Protons and neutrons have about the same mass ($1.67 \cdot 10^{-27}$ kg), however they are some 2000 times heavier than an electron: the nucleus thus contains virtually all of the atom's mass. The nucleus has a density of some 10^{13} g·cm^{-3}.

In order to estimate an atom's mass, since nucleons all have about the same mass, it is thus sufficient to know that atom's **number of nucleons – noted A – also known as its mass number.**

Every one of these particles – i.e. the erstwhile so-called "fundamental" particles – is bound to the atom by a **binding force**; the binding energy, for a particle, being the energy that must be provided to extract it from the atom.

Of the three particles that stand as constituents of the atom, the neutron is the only one that bears no electric charge – hence its name. A proton bears a positive charge, of $+1.6 \cdot 10^{-19}$ C, while an electron bears a negative charge, of $-1.6 \cdot 10^{-19}$ C. This quantity, noted e, is known as the "elementary charge."

Since matter is electrically neutral, an atom thus holds as many protons as it does electrons.

Further information

In 1911, New Zealand-born British physicist Ernest Rutherford was investigating the structure of matter. He was seeking, in particular, to ascertain more precisely the positions of atoms, relative to one another, in matter. He developed a novel model – the so-called Rutherford theory of the atomic nucleus – soon complemented by the model devised by Danish physicist Niels Bohr, in 1913. In this model, atoms consist of a nucleus, of vanishingly small size, compared to the atom as a whole, which nucleus nevertheless holds nearly all of the atom's mass. This nucleus is surrounded by an electron cloud. Electrons travel along stationary orbits, orbiting the nucleus somewhat in the manner of planets around the Sun, as shown in Figure 1.1. Hence the term "planetary model," which is often used to refer to the Bohr model. That model's specific feature was that it allowed for the application of the energy-quantum theory. Electrons may "jump" from one orbit to another, by gaining, or losing a quantum (i.e. a definite, discrete amount) of energy. With the advent of modern quantum theory, this model is now known to be inaccurate.

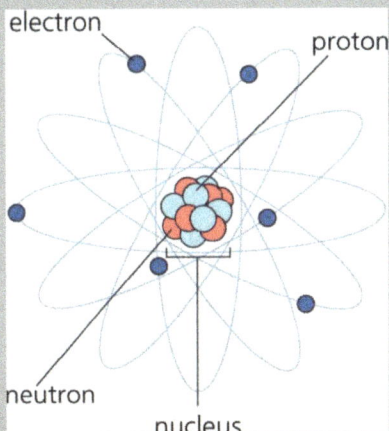

Figure 1.1. Representation of the atom according to the Niels Bohr model.

In 1927, the model put forward by Erwin Schrödinger provided further support for the presence of the nucleus, and its composition, while disallowing the notion of "paths" followed by electrons. It is only possible to determine the region in space where electrons are most frequently to be found: in other words, it is possible to determine the probability of an electron's presence, within a region extending around the nucleus. The radius of the atom now becomes the radius of the region of highest probability for the presence of electrons, around the nucleus. This model is still current, and is shown in Figure 1.2. In this representation, the three electrons of a lithium atom are most probably to be found within the darker regions.

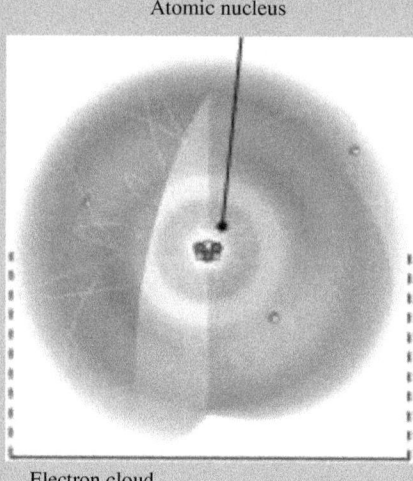

Atomic nucleus

Electron cloud

Figure 1.2. Representation of the atom according to the quantum model.
http://www.cea.fr/Fr/jeunes/livret/Atome/img/Nuage_atome_01.jpg

1.1.2. Definitions. Nomenclature

A **chemical element** (or, more simply, an **element**) is the ensemble of all atoms having nuclei that contain the same number of protons. This number is known as the element's **atomic number**. Atoms of a given element thus all feature – when in the electrically neutral state – the same number of orbiting electrons.

Atoms from one and the same chemical element exhibit identical chemical properties; since they feature the same number of electrons. Indeed, an atom's chemical properties are largely related to the electronic bonds it is able to set up with electrons in neighboring atoms. The differences arising between atoms of one and the same chemical element solely concern the number of neutrons they hold.

Thus, every chemical element has its own name, and is assigned its own symbol, consisting of one or two letters (H for hydrogen, Fe for iron …), together with its atomic number – noted Z – corresponding to the number of protons in the nucleus. Any atom, taken at random from the vast number of atoms, whether presently in existence, or liable to be generated, belongs to a particular "family" of atoms, referred to as a chemical element.

A total of 118 chemical elements have been identified, to date, of which 112 have been given a name, and 89 occur naturally. The elements are often set out in table form, known as the "periodic table" – first devised by Dmitri Mendeleev – an example of which is shown as Annex 1.

To provide a complete description of an atom, the following notation is used:

$^{A}_{Z}X$

- X stands for the element's chemical symbol;
- A, Z indicate, respectively, the number of nucleons, and the number of protons.

A is known as the mass number, Z as the atomic number (also referred to as the charge number). Setting N to stand for the number of neutrons, the relation between these three numbers is:

$$A = N + Z$$

Example:

$^{208}_{82}Pb$

Pb stands for the chemical element lead [Latin *plumbum*]; the atom's mass number is 208, its atomic number is 82; its number of neutrons therefore stands equal to 126.

Hereafter, the Z value will no longer be indicated, since this number is implicitly specified, once the chemical symbol is given. The following notation is therefore used:

^{A}X

Examples: ^{12}C, ^{32}P, ^{56}Fe, ^{60}Co, ^{131}I, ^{222}Rn, ^{238}U, …

1.1.3. Isotopes and isobars

Atoms that are different, while belonging to one and the same chemical element, are known as **isotopes** of that element. Every isotope of a given element thus features the same number of protons (i.e. an identical atomic number Z). For any element, all isotopes exhibit identical chemical properties: this being the common character, serving to define the chemical element. However, isotopes do vary in terms of the number of neutrons they hold, and thus have different mass numbers A.

Example: the isotopes of the element hydrogen: ^{1}H, ^{2}H, ^{3}H.

Isobars are atoms that have the same mass number A, but different atomic numbers. Examples of isobars: ^{14}C, ^{14}N, ^{14}O.

Such atoms will not exhibit any common chemical property.

1.2. Nuclear stability and instability

Nuclei may be grouped into two classes: stable nuclei, involving infinite (or near-infinite) lifetimes; and unstable nuclei, featuring lifetimes ranging from one nanosecond to billions of years.

1.2.1. Stable nuclei

The nuclei of stable atoms feature a number of protons, and of neutrons such that their array remains in perfect equilibrium, and – barring any outside perturbation – this structure does not undergo any alteration.

Examples of stable atoms: ^{1}H, ^{12}C, ^{16}O, ^{39}K, ^{56}Fe, ^{127}I, ...

In Figure 1.3, in which mass number A is plotted along the y-axis, and atomic number Z along the x-axis, the set of all 3 139 nuclei identified to date is shown. Of these nuclei, 256 occur in nature, the others being artificially obtained.

The region marked out in black in the figure plots the distribution of stable nuclei.

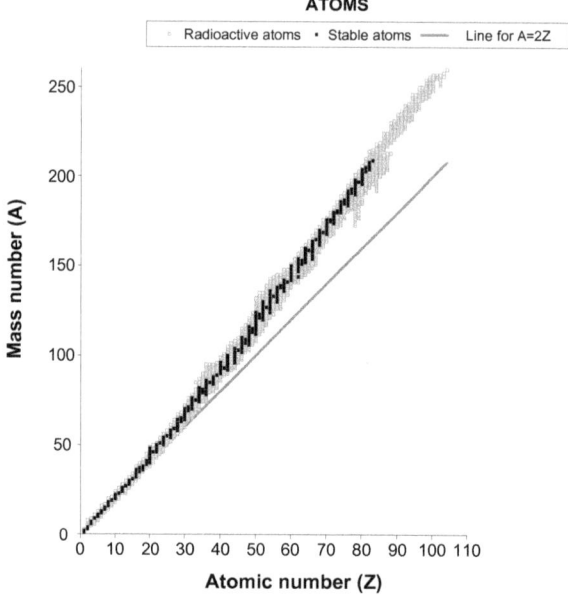

Figure 1.3. Distribution of stable, and radioactive nuclei, as a function of Z, and A.

Stability is found to be the case:

– for light atoms, when they feature a number of protons equal to the number of neutrons;

Examples: ^{4}He, ^{12}C, ^{40}Ca

– for heavier atoms, when they are neutron-rich ($A \approx 2.6Z$). Indeed, the region shown in black, in Figure 1.3, is seen to rise above the line plotting $A = 2Z$.

Examples: ^{133}Cs, ^{180}W, ^{196}Hg, ^{208}Pb

1.2.2. Radioactive nuclei

A nucleus will be stable only if the assembly of Z protons, and N neutrons that make it up is tightly bound (i.e. if the binding energies for its constituent nucleons are high). Should this not be the case, the nucleus is unstable. Such instability shows up as excess energy (and insufficient nucleon binding energies). The nucleus spontaneously transforms into other nuclei, in order to attain a more stable state. This transformation is accompanied by the emission of various kinds of radiation, carrying the energy released.

Such unstable atoms are said to be **radioactive**; they are also known as **radionuclides, or radioisotopes**.

Such atoms may occur naturally, or be produced artificially around particle accelerators (see the chapter on "Radiation of electrical origin").

Examples:

Natural radioactive atoms: ^{14}C, ^{232}Th, ^{238}U
Artificial radioactive atoms: ^{18}F, ^{90}Sr, ^{192}Ir

Of the 118 currently known elements, some thirty do not have a single stable isotope.

Finally, a reminder concerning a definition that is too often disregarded: while, as seen above, a **radionuclide** is a radioactive isotope (e.g. ^{60}Co, ^{137}Cs), a **radioelement** is an element all isotopes of which are radioactive. Technetium (Tc), neptunium (Np), plutonium (Pu) are all radioelements. Thus, it is incorrect to use the term "radioelement" to refer to ^{60}Co, or ^{137}Cs, as is all too often found in the literature.

1.3. Radiation: energy and emission intensity

In general terms, radiation is defined as a propagation mode of energy, traveling across space. A given radiation is thus characterized by its energy level, and emission intensity.

1.3.1. Radiation energy

In the International System of Units (SI), energy is expressed in joules (J). In atomic and nuclear physics, the unit generally used is the electron volt (eV), and multiples of that unit. 1 electron volt corresponds to the energy gained by an electron subjected to a potential difference of 1 volt:

$1 \text{ eV} = 1.6 \cdot 10^{-19}$ J

The following multiples of the electron volt are commonly used:

– the kiloelectron volt (keV): $1 \text{ keV} = 10^3 \text{ eV}$;

– the megaelectron volt (MeV): $1 \text{ MeV} = 10^6 \text{ eV}$;

– the gigaelectron volt (GeV): $1 \text{ GeV} = 10^9 \text{ eV}$.

1.3.2. Radiation: emission intensity

Emission intensity is defined as the percentage of radiation of a specific type, and energy, that is emitted, over the entire set of transformations involved.

By way of example, for 100 emissions of radiation yielded by the decay of a given radioactive atom, should the observed count indicate 80 emissions of type R_1, 15 of type R_2, 5 of type R_3, then the respective emission intensities, for these types of radiation, will stand equal to: $I_{R_1} = 80\%$, $I_{R_2} = 15\%$, $I_{R_3} = 5\%$.

The distribution plotting emission intensity, for a given radiation, as a function of energy, is known as that radiation's "energy spectrum." In the paragraphs that follow, the distinction will be made between line spectra, and continuous energy spectra.

1.4. Nuclear transformation modes

A radionuclide, featuring as it does an unstable nucleus, will tend to stability by undergoing transformation, through **disintegration**, or **deexcitation**. In the disintegration case, the structure of the nucleus is altered, whereas, in a deexcitation mode, excess energy is released, while the structure remains unaltered. The generally adopted terminology, however, refers to both cases as **decay** modes.

Radionuclides are sources of different kinds of radiation, respectively identified by the first three letters of the Greek alphabet: alpha (α), beta (β), gamma (γ). They have been found to consist, respectively, of helium-4 nuclei, electrons, and photons.

Every radioactive transformation complies with – among other physical laws – the general physical laws of conservation of electric charge, conservation of momentum, and conservation of total energy.

1.4.1. Radioactive disintegration modes

Radioactive decay involving disintegration yields a daughter atom with an atomic number Z that differs from that of the parent atom – hence one belonging to a different chemical element. This is a transmutation phenomenon.

1.4.1.1. Alpha decay

As may be seen from Figure 1.3, all isotopes of heavy chemical elements, i.e. of elements featuring a Z value higher than 83, are radioactive. If they are to tend to stability, such isotopes must lower their mass. Within the radioactive atom's nucleus, two protons and two neutrons come together to form a helium-4 (^{4}He) nucleus, which is expelled, this constituting what is known as α radiation. α decay may be represented by way of the following relations:

$$^A_Z X_N \rightarrow {}^{A-4}_{Z-2} Y_{N-2} + {}^4_2 He_2$$

or:

$$^A X \rightarrow {}^{A-4} Y + \alpha$$

This decay mode by disintegration results in a new element, with an atomic number Z lower by 2 units, compared to the parent element.

It may be noted that before, and after this decay process, the total mass number A, and total electric charge (i.e. the total number of electrons, and protons) are both conserved.

1 – Radioactivity and nuclear physics

All spontaneous disintegration phenomena yield energy. They are thus termed **exoergic** processes. The equivalence of mass, and energy is a well-known phenomenon, as set out in the relation: $E = mc^2$. Since radioactive transformations release energy, mass variations are found to arise between parent, and daughter nuclei. Consequently, no disintegration can occur whereby the mass of the parent nucleus would be lower than the daughter nucleus (or the combined masses of the daughter nuclei). This excess energy, which may be calculated by way of the mass difference found for the decay products, is known as the disintegration energy (also referred to as decay heat), and noted Q (hence, it is commonly referred to as the Q-value for that process).

In the alpha-decay case, the energy conservation law, for nuclear masses (i.e. for the masses of the nuclei involved), during this transformation, may be written as follows:

$$Q = m_X c^2 - (m_Y c^2 + m_\alpha c^2)$$

where c is the velocity of light, expressed in meters per second (m·s^{-1}), m the nuclear mass, expressed in kilograms (kg), respectively for the parent, and daughter nuclei, and for the alpha particle. The energy thus calculated is expressed in joules (J).

As indicated earlier, the energies involved, in nuclear physics, are very much smaller than the joule, and values are converted to electron volts (eV), multiples of that unit (keV, MeV) being commonly used for that purpose.

When tables are used, atomic masses are more readily found. Hence, by adding to both terms, in the above equation, the mass-energy for Z electrons, while disregarding differences in electron binding energies, the energy balance may be written as a function of the atomic masses involved:

$$Q = M_X c^2 - (M_Y c^2 + M_\alpha c^2)$$

The available energy, Q, does not vanish away: it takes the form of kinetic energy, imparted to the decay products. The alpha particle, being lighter than the daughter nucleus, Y, carries off almost all of the available energy (about 98%). The remaining energy is taken up by the daughter nucleus. Compliance with the laws of conservation of momentum, and conservation of kinetic energy entails that:

$$E_\alpha = \left(\frac{A-4}{A}\right) Q$$

where E_α is the kinetic energy of the α particle, A is the mass number for the parent nucleus.

If the alpha radiation carries off all of the excess energy present in the parent nucleus, then the daughter nucleus is generated in its so-called "ground" state. Such a radionuclide is known as a "pure" emitter.

Example:

$$^{210}\text{Po} \rightarrow {}^{206}\text{Pb} + \alpha$$

Using tables listing the atomic masses for the various decay products, the available energy may be calculated. This is found to be:

$$Q = M(^{210}\text{Po})c^2 - M(^{206}\text{Pb})c^2 - M(^{4}\text{He})c^2 = 5407 \text{ keV}$$

As previously pointed out, the alpha radiation carries off a large proportion of this energy:

$$E_\alpha = [(A-4)/A] \times Q = (206/210) \times 5407 = 5304 \text{ keV} \qquad I_\alpha = 100\%$$

To sum up: 100% of polonium-210 decay processes yield the ground state of its daughter product, lead-206, with a concomitant emission of alpha radiation, with an energy of 5.304 MeV.

When some energy is left to the daughter nucleus, the latter is said to be produced in an excited state, this being noted $^{A-4}Y^*$. If the daughter nucleus, Y, is generated in an excited state, with an energy E^* relative to the ground state value, then the total available energy comes down, to the value $Q - E^*$. Consequently, the alpha particle then carries less energy.

Example:

$$^{232}\text{Th} \rightarrow {}^{228}\text{Ra}^* + \alpha$$

Using tables listing the atomic masses for the various decay products, the available energy may be calculated. This is found to be:

$$Q = M(^{232}\text{Th})c^2 - M(^{228}\text{Ra})c^2 - M(^4\text{He})c^2 = 4083 \text{ keV}$$

Alpha radiation carries off a large proportion of this energy:

$$E_{\alpha_1} = [(A-4)/A] \times Q = (228/232) \times 4083 = 4013 \text{ keV} \qquad I_{\alpha_1} = 77\%$$

Thorium-232 decay processes sometimes yield the daughter nucleus in an excited state, at 60 keV (in 23% of cases, according to experimental data). Consequently, the alpha particle's kinetic energy is then lower:

$$E_{\alpha_2} = [(A-4)/A] \times (Q - E^*) = (228/232) \times (4083 - 60) = 3954 \text{ keV} \qquad I_{\alpha_2} = 23\%$$

77% of thorium-232 decay processes yield the ground state of its daughter product, radium-228, together with emission of α radiation, with an energy of 4.01 MeV; the remaining 23% yield the daughter nucleus in an excited state, the α particles emitted then carrying less energy.

In either case, the alpha radiation involved is monoenergetic: i.e. it carries a single, discrete energy level. The energy spectrum for the α particles is a line spectrum, the lines as a rule lying quite close to one another, as shown in Figure 1.4.

As a whole, emission energy values, for alpha radiation yielded by radioactivity, lie in the 3–9 MeV range.

1.4.1.2. Beta-minus decay

As may be noted in Figure 1.3, neutron-rich nuclides are radioactive. In order to come closer to stability, one neutron, within the nucleus, transforms into a proton. Conservation of electric charge, as such a decay takes place, entails that, alongside the proton, one negatively charged electron – also known as a negatron – must be generated: this is then referred to as β^- radiation. Conservation laws governing other quantities – which need not

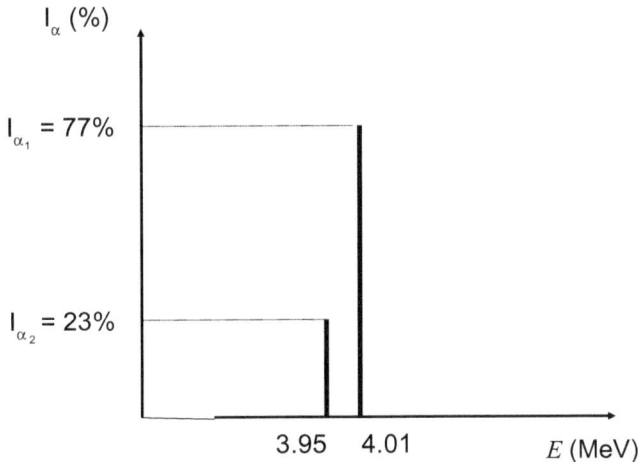

Figure 1.4. Energy spectrum of alpha particles emitted by thorium-232.

be further specified in the present context – result in an electrically neutral particle also being emitted, having an infinitely small mass: an antineutrino.

β^- decay may be represented by way of the following relations:

$$^{A}_{Z}X_N \to ^{A}_{Z+1}Y_{N-1} + \beta^- + \bar{\nu}_e$$

or:

$$^{A}X \to ^{A}Y + \beta^- + \bar{\nu}_e$$

or:

$$n \to p + e^- + \bar{\nu}_e$$

This decay mode, taking the form of the transformation of a neutron into a proton, allows atoms that are unstable because they are neutron-rich – i.e. owing to an overlarge number of neutrons – to attain a more stable state. It should be noted that both parent, and daughter nuclei exhibit the same mass number A: this is an isobaric transformation.

Such a transformation yields a new element, having an atomic number Z that is higher by one unit than that of the parent element.

It should again be pointed out that both mass number A, and total electric charge are conserved, before and after decay. The conservation of nuclear mass-energies, during this transformation, may be written as follows:

$$Q = m_X c^2 - (m_Y c^2 + m_e c^2 + m_{\bar{\nu}} c^2)$$

The mass of the antineutrino may be regarded as negligible.

If atomic masses are used, the energy balance may now be written as follows:

$$Q = M_X c^2 - M_Y c^2$$

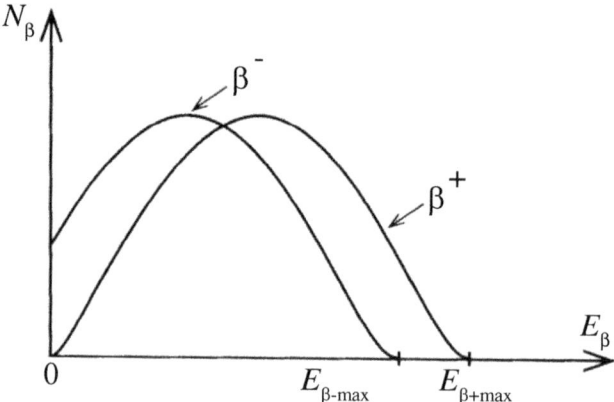

Figure 1.5. Energy spectra for β emissions.

Bearing in mind the low masses of the β^- and $\bar{\nu}$ particles emitted, the recoil kinetic energy of the daughter product Y may be disregarded, and all of the available energy Q takes the form of kinetic energy imparted to the beta–antineutrino pair.

Owing to its characteristic properties, the antineutrino may not be detected by conventional means. Only the β^- radiation can be observed, and interacts with matter (see Chapter 2, "Interaction of radiation with matter").

In every decay process, as previously mentioned, the total energy released, E, is carried by the decay products, being distributed, in this case, in random fashion between the β^- radiation and the antineutrino emitted:

$$E = E_{\beta^-} + E_{\bar{\nu}_e}$$

This results in the energy emission spectrum, for β^- radiation, taking the form of a continuous spectrum, ranging from zero energy (all of the energy released is delivered to the antineutrino, $E = E_{\bar{\nu}_e}$) though to a maximum energy value (all of the energy released is delivered to the β^- radiation, $E = E_{\beta^- \max}$).

The general shape for curves plotting β^- emission spectra is shown in Figure 1.5. The precise shape of the spectrum curve is characteristic for a given emitter radionuclide.

The maximum energy value is the sole indication provided by publications setting out characteristic quantities for radionuclides. As a whole, it may be noted that maximum emission energy values, for β^- radiation yielded by radioactivity, lie in the 10 keV–3 MeV range.

It may be useful, in some cases, to consider the average energy, for this continuous distribution. The following rule-of-thumb formula allows the average energy to be derived from the maximum energy value: $E_{av} = \frac{E_{\beta \max}}{3}$.

Example:

$$^{32}P \rightarrow {}^{32}S + \beta^- + \bar{\nu}_e$$

From tables listing the atomic masses for the various decay products, the available energy may be calculated. This is found to be:

$$Q = M(^{32}P)c^2 - M(^{32}S)c^2 = 1710 \text{ keV}$$

In like manner to one of the cases described for the previous decay mode, the beta-minus particle and the antineutrino carry off all of the available energy:

$$E_{\beta^- \text{ max}} = 1.71 \text{ MeV} \quad I_{\beta^- \text{ max}} = 100\%$$

This is an instance of a pure β^--emitter.

As was also the case for α decay processes, on the other hand, some β^- decay processes yield an excited state of the daughter nuclide:

$$^{137}\text{Cs} \rightarrow {}^{137}\text{Ba} + \beta^- + \overline{\nu}_e$$

$$E_{\beta^- \text{ max}_1} = 1.17 \text{ MeV} \quad I_{\beta^- \text{ max}_1} = 5\%$$
$$E_{\beta^- \text{ max}_2} = 0.51 \text{ MeV} \quad I_{\beta^- \text{ max}_2} = 95\%$$

1.4.1.3. Beta-plus decay

Just as is the case with β^- decay, proton-rich nuclei are equally radioactive. In order to come closer to stability, one proton, within the nucleus, transforms into a neutron. Conservation of electric charge, as such a decay takes place, entails that, alongside the neutron, one positively charged electron – known as a positron – must be generated: this is then referred to as β^+ radiation. This is likewise found to involve the emission of a neutrino.

β^+ decay may be represented by way of the following relations:

$$^A_Z X_N \rightarrow {}^A_{Z-1} Y_{N+1} + \beta^+ + \nu_e$$

or:

$$^A X \rightarrow {}^A Y + \beta^+ + \nu_e$$

or:

$$p \rightarrow n + e^+ + \nu_e$$

This decay mode, taking the form of the transformation of a proton into a neutron, allows proton-rich atoms to attain a more stable state. As is the case with β^- decay, this is an isobaric transformation, which yields a new element having an atomic number Z that is lower by one unit than that of the parent element.

The conservation of nuclear mass-energies, during this transformation, may be written as follows:

$$Q = m_X c^2 - (m_Y c^2 + m_{e^+} c^2 + m_\nu c^2)$$

The mass of the neutrino is negligible.

If atomic masses are used, the energy balance may now be written as follows:

$$Q = (M_X c^2 - M_Y c^2) - 2 m_{e^-} c^2$$

For β^+ decay to be possible, in energy terms, the difference between atomic masses, for the parent, and daughter atoms, must be at least equal to $2 m_{e^-}$. The mass-energy of an electron, as shown by the following calculation, standing equal to:

$E = m_{e^-} c^2 = 9.11 \cdot 10^{-31}$ kg $\times (299\ 792\ 458$ m/s$)^2 = 8.18765968 \cdot 10^{-14}$ J   i.e. 511 keV

The mass-energy difference, between parent, and daughter atoms, must be larger than 1022 keV.

As in the case of beta-minus decay, owing to the low masses of the β^+ and ν particles emitted, the recoil kinetic energy of the daughter product Y may be disregarded, and all of the available energy Q takes the form of kinetic energy imparted to the beta–neutrino pair.

In every such decay process, as in the previous case, the total energy released, E, is distributed in random fashion between the β^+ radiation and the neutrino emitted:

$$E = E_{\beta^+} + E_{\nu_e}$$

This results in the energy emission spectrum, for β^+ radiation, likewise taking the form of a continuous spectrum (see Figure 1.5). The offset observed, for β^+ spectra, towards higher energies is due to the Coulomb repulsion affecting β^+ radiation, from the constituent protons in the nucleus.

Maximum emission energy values, for β^+ radiation yielded by radioactivity, may be said to lie in the 10 keV–3 MeV range, average energy being approximately equal to one third of maximum energy $E_{\beta^+ \text{max}}$.

Example:

$$^{13}\text{N} \rightarrow {}^{13}\text{C} + \beta^+ + \nu_e$$

From tables listing the atomic masses for the various decay products, the available energy may be calculated. This is found to be:

$$Q = M(^{13}\text{N})c^2 - M(^{13}\text{C})c^2 - 2m_{e^-}c^2 = 1198 \text{ keV}$$

As in previous cases, the beta-plus particle and the neutrino carry off all of the available energy:

$$E_{\beta^+ \text{max}} = 1.20 \text{ MeV} \quad I_{\beta^+ \text{max}} = 100\% \quad (\text{pure } \beta^+ \text{ emitter})$$

1.4.1.4. Electron capture

As was pointed out earlier, proton-rich nuclides are radioactive, and are able to tend to stability by way of β^+ decay. However, another, competing phenomenon also occurs: electron capture.

This process corresponds to a reaction whereby a proton-rich nucleus captures an electron from the atom's electron cloud. One proton then combines with the electron, to yield a neutron, which remains within the nucleus. This is also found to involve the emission of a neutrino.

Electron-capture decay may be represented by way of the following relations:

$$^A_Z\text{X} + e^- \rightarrow {}^A_{Z-1}\text{Y} + \nu_e$$

or:

$$p + e^- \rightarrow n + \nu_e$$

This decay mode yields a new element having an atomic number Z that is lower by one unit than that of the parent element.

Whether through β^+ decay, or electron capture, the selfsame daughter product is obtained.

The nucleus preferentially captures an electron lying on a so-called core orbit, i.e. an orbit that comes very close to the nucleus. This phenomenon disturbs the strict organization of the electron cloud, which then rearranges itself, to expel, from one orbit to the next, the vacancy thus arising, to the atom's outer reaches. This rearrangement gives rise to the emission of electromagnetic radiation (X-rays), and/or electron emissions (so-called "Auger electrons:" see section 5.2.2). The energy spectra obtained, for these radiation emissions, are line spectra.

The energy values, for such radiation of extra-nuclear origin, are related to the atom's atomic number Z, though they are invariably found to stand lower than 100 keV.

The following empirical relation allows an approximate value to be arrived at, for the X-ray emission energy: $E_X(\text{keV}) = \frac{Z^2}{100}$.

The conservation of nuclear mass-energies (i.e. for the mass-energies of the nuclei involved), during this transformation, may be written as follows:

$$Q = m_{e^-}c^2 + m_X c^2 - (m_Y c^2 + m_\nu c^2)$$

The mass of the neutrino may be regarded as negligible.

If atomic masses are used, the energy balance may now be written as follows:

$$Q = M_X c^2 - M_Y c^2$$

This equation shows that electron-capture decay is possible whenever the parent atom's atomic mass is such that: $M_X > M_Y$. Whereas the competing β^+ decay mode is possible, in energy terms, only in those cases where: $M_X > M_Y + 2m_{e^-}$.

Example:

$$^{55}\text{Fe} + e^- \rightarrow {}^{55}\text{Mn} + \nu_e$$

^{55}Fe: $I_{EC} = 100\%$

From tables listing the atomic masses for the various decay products, the available energy may be calculated:

$$Q_{EC} = M(^{55}\text{Fe})c^2 - M(^{55}\text{Mn})c^2 = 231 \text{ keV}$$

The neutrino carries off all of the available energy. Consequently, no radiation of intra-nuclear origin can be detected. On the other hand, 5.9-keV X-rays, yielded by the electron cloud rearrangement, may be detected.

Proton-rich atoms thus fall into three categories:

– those that are liable to undergo β^+ decay only;

– those that are liable to undergo only electron capture (EC) decay;

– those that may decay either through electron capture, or β^+ emission.

In the latter case, it is necessary to ascertain the respective percentages of decay processes for either mode. This is dependent on two factors: the mass of the radioactive nucleus, and the excess energy carried by that nucleus. The heavier the nucleus, the more prone it is to decay by way of electron capture. On the other hand, the larger the amount of excess energy residing in the nucleus, the more liable that nucleus is to decay by β^+ emission. Indeed, it can be shown that excess energy, within the nucleus, must be at least equal to 1.022 MeV for β^+ decay to occur.

Example:

$$^{18}F \rightarrow {}^{18}O + \beta^+ + \nu_e$$

$$^{18}F + e^- \rightarrow {}^{18}O + \nu_e$$

$E_{\beta^+ \, max} = 0.63$ MeV $I_{\beta^+ \, max} = 97\%$; therefore: $I_{EC} = 3\%$.

Experimental data show 97% beta-plus decay events, and 3% electron-capture decay processes.

Using tables listing the atomic masses for the various decay products, the available energies may be calculated, for both the beta-plus decay case, and the electron capture case:

$$Q_{\beta^+} = M(^{18}F)c^2 - M(^{18}O)c^2 - 1022 = 633 \text{ keV}$$

$$Q_{EC} = M(^{18}F)c^2 - M(^{18}O)c^2 = 1655 \text{ keV}$$

This source yields both β^+ radiation, with a maximum energy of 633 keV, and X-photons during electron capture events.

From the definition of emission intensity, it will be seen that the sum of emission intensities, for all types of nuclear radiation emitted in disintegration processes (α, β^-, β^+, EC), must be equal to 100%.

1.4.2. Gamma decay

Subsequent to a disintegration process, the daughter nucleus may be left in an excited state. Except in very specific cases – so-called metastable nuclei (see section 4.3) – such a nucleus instantly undergoes deexcitation, as a rule through emission of a kind of electromagnetic radiation known as γ radiation.

Such nuclear deexcitation may be represented by way of the following relation:

$$^AY^* \rightarrow {}^AY + \gamma$$

where ^AY stands for the ground state of nucleus Y, or some excited state of that nucleus, of energy lower than that of $^AY^*$.

Example:

^{95}Nb: $E_{\beta^- \, max} = 0.16$ MeV $I_{\beta^- \, max} = 99\%$

$E_\gamma = 0.77$ MeV $I_\gamma = 99\%$

Coming down to the ground state may involve, in some cases, going through a number of intermediate excited states: this results in the emission of several γ rays. For some radionuclides, owing to this cascade phenomenon, the sum total of emission intensities, for all of the individual γ rays emitted, may come to more than 100%.

Example:

^{60}Co: $E_{\beta^-\text{max}} = 0.32$ MeV $I_{\beta^-\text{max}} \approx 100\%$

$E_{\gamma_1} = 1.17$ MeV $I_{\gamma_1} = 100\%$

$E_{\gamma_2} = 1.33$ MeV $I_{\gamma_2} = 100\%$

Energy spectra, for γ radiation, take the form of line spectra.

Emission energies, for γ radiation yielded by radioactivity, as a rule range from 60 keV to 3 MeV.

Further information

Internal conversion electrons

As we have just seen, coming down to the ground state, for a daughter nucleus that has been emitted in an excited state, involves a release of energy. This energy may either be emitted in the form of γ electromagnetic radiation, or be transferred to an electron in the atom's electron cloud, which is then ejected from the atom. Such an electron is known as an "internal conversion electron".

The energy of an internal conversion electron is equal to the difference between the energy of the corresponding γ radiation, and the energy required to strip the electron from the electron cloud (i.e., that electron's binding energy). Thus, for instance, an internal conversion electron originating in the K-shell will have an energy equal to E_γ, minus the K-shell binding energy:

$$E_{ce_K} = E_\gamma - E_{b_K}$$

Owing to the vacancy generated in the electron cloud, this causes a rearrangement of the cloud, resulting in emission of X-radiation, identical, in kind and nature, to that described for the electron capture decay case.

1.4.3. Metastable nuclei

In a fairly small number of instances, the daughter nucleus, when emitted in an excited state, does not deexcite immediately, rather it does so with a half-life that is specific to it. Such specific excited states are referred to as "metastable levels."

The best-known instance is that of metastable technetium-99 – noted ^{99m}Tc – which is very widely used in nuclear medicine (see Chapter 6, "Uses of sources ionising radiation"). This is obtained from molybdenum-99 via the following reactions:

$$^{99}Mo \rightarrow {}^{99m}Tc \rightarrow {}^{99}Tc$$

Molybdenum-99 has a half-life of 66 hours, technetium-99m a half-life of 6 hours

Such metastable levels are to be considered as exhibiting a specific, independent decay mode, by deexcitation, if their half-life exceeds 1 nanosecond (ICRP Publication 107).

1.5. The electron cloud

1.5.1. Configuration of an electron cloud

Let us take a further look at the atom, with its nucleus at the center, and its electron cloud. The electron cloud exhibits a highly structured configuration. Its constituent electrons are distributed, by energy, into shells. Each shell is characterized by a particular binding energy, and a maximum capacity. The binding energy corresponds to the minimum energy that must be imparted to an electron to remove it from its shell. The shell's capacity, i.e. the number of electrons it can accommodate, corresponds to the maximum number of electrons that may have that particular binding energy (see Table 1.1).

Table 1.1. Electron shell nomenclature.

SHELLS	K	L	M	N	O ...
Shell number: n	1	2	3	4	5 ...
Capacity	2	8	18	32	50
Binding energy	E_K	E_L	E_M	E_N	E_O

The configuration of an electron cloud complies with a number of rules, the most important of which are the following ones:

– the most strongly bound shells are completely filled;

– only the three least bound shells – at most – may remain partly unfilled;

– the least bound shell may not be occupied by more than 8 electrons.

The following table sets out, for a number of elements, the binding energies for the three most strongly bound shells. As regards the other shells, the corresponding binding energies tend to be negligible, for the purposes of the present account. A general finding is that the heavier the element, the higher the binding energies that arise.

In actual fact, the atom's electron cloud is made up of shells, but also of subshells, which we have not sought to describe, in order to keep matters simple. The energy values listed in Table 1.2 are thus average values, over the various subshells involved.

Table 1.2. Average binding energies for atomic electrons (keV).

Element		Z	Shells		
			K	L	M
Hydrogen	H	1	0.013		
Helium	He	2	0.027		
Lithium	Li	3	0.055		
Beryllium	Be	4	0.111		
Boron	B	5	0.19		
Carbon	C	6	0.28		
Nitrogen	N	7	0.40		
Oxygen	O	8	0.53		
Fluorine	F	9	0.68		
Neon	Ne	10	0.87	0.02	
Sodium	Na	11	1.06	0.03	
Magnesium	Mg	12	1.30	0.05	
Aluminum	Al	13	1.56	0.08	
Silicon	Si	14	1.84	0.10	
Phosphorus	P	15	2.14	0.13	
Sulfur	S	16	2.47	0.17	
Chlorine	Cl	17	2.82	0.21	
Argon	Ar	18	3.20	0.27	
Potassium	K	19	3.61	0.31	
Calcium	Ca	20	4.04	0.36	
Scandium	Sc	21	4.50	0.42	
Titanium	Ti	22	4.97	0.47	
Vanadium	Va	23	5.47	0.53	
Chromium	Cr	24	5.99	0.60	
Manganese	Mn	25	6.54	0.67	
Iron	Fe	26	7.11	0.74	
Cobalt	Co	27	7.71	0.82	
Nickel	Ni	28	8.33	0.90	
Copper	Cu	29	8.98	0.98	0.04
Zinc	Zn	30	9.66	1.07	0.05
Gallium	Ga	31	10.37	1.17	0.06
Germanium	Ge	32	11.10	1.27	0.08
Arsenic	As	33	11.86	1.38	0.09
Selenium	Se	34	12.65	1.50	0.11
Bromine	Br	35	13.48	1.62	0.13

(continued on next page)

Table 1.2. (Continued)

Element		Z	Shells		
			K	L	M
Krypton	Kr	36	14.32	1.74	0.15
Rubidium	Rb	37	15.20	1.89	0.18
Strontium	Sr	38	16.11	2.03	0.21
Yttrium	Y	39	17.04	2.17	0.23
Zirconium	Zr	40	18.00	2.32	0.26
Niobium	Nb	41	18.99	2.48	0.29
Molybdenum	Mo	42	20.00	2.63	0.32
Technetium	Tc	43	21.05	2.80	0.35
Ruthenium	Ru	44	22.12	2.97	0.38
Rhodium	Rh	45	23.22	3.14	0.41
Palladium	Pd	46	24.35	3.32	0.44
Silver	Ag	47	25.51	3.51	0.48
Cadmium	Cd	48	26.71	3.71	0.52
Indium	In	49	27.94	3.91	0.56
Tin	Sn	50	29.20	4.12	0.61
Antimony	Sb	51	30.49	4.34	0.65
Tellurium	Te	52	31.81	4.56	0.71
Iodine	I	53	33.17	4.79	0.77
Xenon	Xe	54	34.59	5.03	0.82
Cesium	Cs	55	35.98	5.27	0.88
Barium	Ba	56	37.44	5.53	0.94
Lanthanum	La	57	38.93	5.79	1.01
Cerium	Ce	58	40.45	6.04	1.06
Praseodymium	Pr	59	42.00	6.30	1.12
Neodymium	Nd	60	43.57	6.57	1.17
Promethium	Pm	61	45.19	6.84	1.23
Samarium	Sm	62	46.84	7.12	1.28
Europium	Eu	63	48.52	7.41	1.35
Gadolinium	Gd	64	50.22	7.70	1.41
Terbium	Tb	65	51.99	8.00	1.47
Dysprosium	Dy	66	53.78	8.30	1.53
Holmium	Ho	67	55.60	8.61	1.59
Erbium	Er	68	57.46	8.93	1.63
Thulium	Tm	69	59.38	9.26	1.74
Ytterbium	Yb	70	61.31	9.59	1.80
Lutetium	Lu	71	63.31	9.92	1.87

(continued on next page)

Table 1.2. (Continued)

Element		Z	Shells		
			K	L	M
Hafnium	Hf	72	65.32	10.28	1.95
Tantalum	Ta	73	67.41	10.64	2.04
Tungsten	W	74	69.52	11.00	2.12
Rhenium	Re	75	71.67	11.39	2.21
Osmium	Os	76	73.87	11.77	2.30
Iridium	Ir	77	76.11	12.16	2.39
Platinum	Pt	78	78.39	12.57	2.48
Gold	Au	79	80.73	12.98	2.59
Mercury	Hg	80	83.12	13.40	2.69
Thallium	Tl	81	85.53	13.84	2.80
Lead	Pb	82	88.01	14.29	2.95
Bismuth	Bi	83	90.54	14.74	3.03
Polonium	Po	84	93.11	15.20	3.14
Astatine	At	85	95.74	15.68	3.26
Radon	Rn	86	98.41	16.16	3.38
Francium	Fr	87	101.14	16.65	3.51
Radium	Ra	88	103.93	17.15	3.63
Actinium	Ac	89	106.76	17.67	3.75
Thorium	Th	90	109.65	18.19	3.89
Protactinium	Pa	91	112.60	18.54	4.03
Uranium	U	92	115.61	19.26	4.15
Neptunium	Np	93	118.66	19.80	4.29
Plutonium	Pu	94	121.77	20.37	4.43
Americium	Am	95	124.94	20.94	4.58
Curium	Cm	96	128.16	21.53	4.72
Berkelium	Bk	97	131.45	22.13	4.88
Californium	Cf	98	134.79	22.74	5.03
Einsteinium	Es	99	138.19	23.35	5.18
Fermium	Fm	100	141.66	23.98	5.34
Mendelevium	Md	101	146.78	24.73	5.49
Nobelium	No	102	150.54	25.39	5.64
Lawrencium	Lr	103	154.38	26.06	5.78
Rutherfordium	Rf	104	158.30	26.67	5.93

Table 1.3. Example: characteristic X-ray lines of tungsten.

X-ray lines	K_{α_1}	K_{α_2}	K_{β_1}	K_{β_2}	L_{α_1}	L_{α_2}
X-ray energies (keV)	59.3	58	67.2	69.1	8.4	8.3

1.5.2. Spontaneous rearrangement

When a nucleus undergoes disintegration through electron capture, or deexcitation by way of internal conversion, electrons – chiefly from the K, and L shells – are removed. The disappearance of an electron in a very strongly bound shell (from the K shell, say) causes a spontaneous rearrangement to take place, from one shell to another, to fill the more strongly bound shells. Electron rearrangement actually occurs from subshell to subshell – aside from the K shell, which has no subshell.

As an electron jumps from one electron shell, characterized by a particular binding energy (E_L, for instance), to another, more strongly bound shell (e.g. E_K), energy is released:

$$\Delta E = E_K - E_L$$

The closer the electron shells involved are to one another, the higher the probability that such phenomena will occur. Thus, it is far more likely that an electron from the L shell, rather than from the M shell, will fill a vacancy arising in the K shell.

This energy release, ΔE, occurs in the form either of an X-ray emission, or of the ejection of a bound electron, known as an Auger electron emission.

1.5.2.1. X-ray emission

X-rays are a form of electromagnetic radiation, similar to gamma radiation. The X-rays emitted correspond to discrete energy transitions: in other words, they have well defined, distinct energies, and a cascade rearrangement gives rise to an energy spectrum for the photons emitted. This takes the form of a spectrum consisting of discrete, separate lines.

If a K-shell electron is replaced by an electron from the L shell, a spectrum line is emitted, known as a K_α line. If the electron reaching the K shell comes from the M shell, or the N shell, the corresponding line is designated K_β. If the vacancy occurs in the L shell, electrons originating in the M, N, O shells will generate, L_α, L_β, L_γ lines. It should be noted that the replacing electrons in fact originate from distinct subshells. To differentiate between the X-ray lines corresponding to these various subshells, the lines are designated K_{α_1}, K_{α_2}, K_{β_1}, K_{β_2}, L_{α_1}, L_{α_2} ...

The resulting spectrum is known as the atom's "characteristic" X-ray spectrum. The X-ray energies making it up, equal as they are to the differences in energy levels between successive shells and subshells, are likewise different from one atom to another, and thus stand as a veritable signature for that element.

Table 1.3 sets out, by way of example, some of the characteristic lines for tungsten ($Z = 74$).

1.5.2.2. Auger electron emission

In the event of an Auger electron emission, the energy released, as the electron cloud undergoes rearrangement, is transferred to another electron in the cloud, belonging to a less strongly bound shell. The latter, known as an Auger electron, is then ejected with the following energy:

$$E_A = \Delta E - E_b = (E_K - E_L) - E_b$$

where E_b is the binding energy of the electron to which energy ΔE is imparted.

The vacancy generated, within the electron cloud, by the ejection of the Auger electron results, here again, in the cloud undergoing rearrangement, with an ensuing further emission of X-rays, or Auger electrons.

1.5.2.3. Fluorescence yield

The two phenomena described above are competing processes (emission of Auger electrons reduces the yield of X-ray emissions). The competition arising between these two processes defines the fluorescence yield ω_S for a particular shell:

$$\omega_S = \frac{\text{number of X-rays emitted}}{\text{number of S-shell vacancies}}$$

$$(S = K, L, M, \ldots)$$

The difference between 1, and the fluorescence yield ω_S gives the Auger electron emission yield for shell S.

For the K and L shells, the respective fluorescence yields are noted ω_K, ω_L. Many tables have been published, setting out the numerical values for ω_K, ω_L. Figure 1.6 plots the ω_K and ω_L fluorescence yields, as a function of the atoms' atomic number Z. Thus, the way ω values change with Z shows that X-ray emissions predominate for heavy elements. By contrast, the Auger effect is significant for light elements.

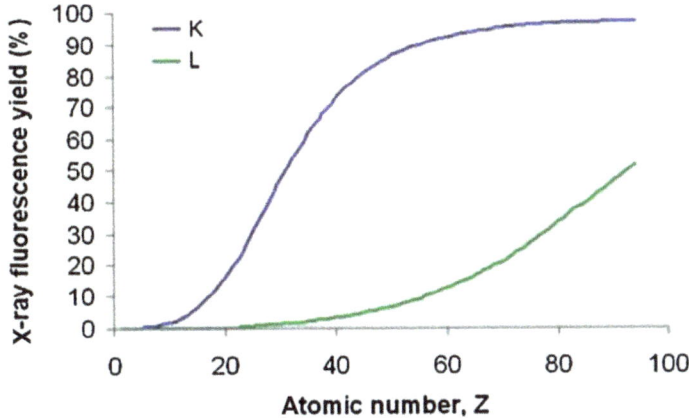

Figure 1.6. Variation of fluorescence yields ω_K (blue), ω_L (green) as a function of atomic number.

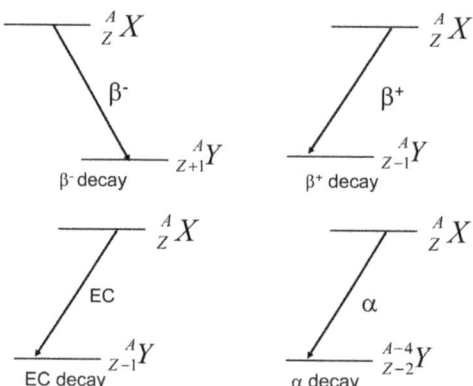

Figure 1.7. The graphic symbols used in decay schemes for disintegration processes.

1.5.3. Induced rearrangement

Aside from spontaneous X-ray emissions, other X-ray emission modes may arise. These are the outcome of interactions occurring between electrons and matter, and photon interactions (see Chapter 2, "Interaction of ionising radiation with matter"). Such interactions may take three different forms, depending on the energy involved, and the nature of the medium the electron travels through:

- interaction of electrons with the nucleus, resulting in bremsstrahlung ("braking radiation") X-radiation;

- interaction of incoming electrons with the electron cloud, causing ionisation, or excitation, directly resulting in a rearrangement of the cloud;

- photon interaction, by way of the photoelectric effect.

1.6. Decay schemes

A decay scheme has the purpose of representing, by means of lines, and arrows, the transformation of a parent nucleus, ^A_ZX, into a daughter nucleus, $^{A'}_{Z'}Y$. The initial state of the parent nucleus corresponds to an energy equivalent that is higher than that found for the final state. This has led to representing the energy levels for the initial, and daughter nuclides as horizontal lines, the latter level corresponding to a line set below the first one (lower energy), and offset to the right, or left, according to whether Z increases, or decreases, as a result of the decay process.

Accordingly, each disintegration mode corresponds to the following decay schemes – not taking into account any possible excited states, for the levels shown.

In many cases, the daughter nucleus is not generated in its ground state. The schemes as shown above are thus complemented by a representation, somewhat like rungs in a ladder, showing the excitation levels obtained, subsequent to the disintegration process.

This ladder takes the form of horizontal lines, indicating the **daughter's energy** states, **in keV**. The bottom line corresponds to the ground state (level of energy 0).

The various decay modes are quantified in terms of the relevant emission intensities, expressed as percentages (%). These intensities express the proportion of initial nuclei that transform according to the path shown. The sum of all intensities, for the various disintegration modes, for a given radionuclide, must always come to 100%.

The deexcitation modes, for the resulting levels, are likewise characterized by emission intensities. In such a case, the sum of all intensities arriving at a given level must always stand equal to the sum total of outgoing intensities from that level. Furthermore, the sum of all emission intensities reaching the ground state level must always stand equal to 100%.

A number of publications are available, that set out the decay schemes for every radioactive nuclide.

By way of example, consider the simplify decay scheme for sodium-22 (see Figure 1.8).

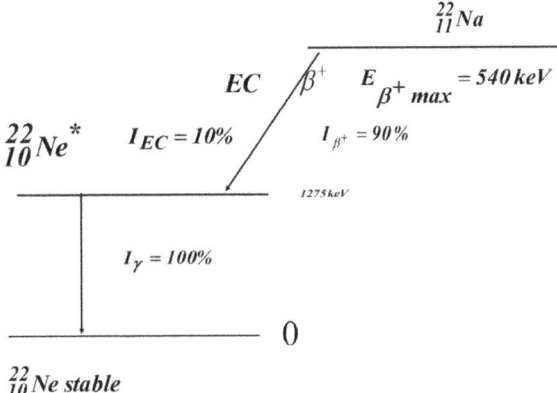

Figure 1.8. The simplify decay scheme for sodium-22.

The usefulness of decay schemes may readily be understood, making it possible as they do, by means of highly expressive symbols, to avoid the need for a description that would prove, in most cases, intractable, if not impossible.

Thus, by looking at the decay scheme for sodium-22, it can be seen that two kinds of disintegration (EC, β^+) both yield the same excited level of neon-22. This excited level deexcites solely by way of a single gamma-ray emission, the intensity for which stands equal to the sum of the intensities for the two disintegration modes that arrive at that level. As a result, as may be seen: $I_\gamma = 100\%$

1.7. Physical quantities and fundamental properties

1.7.1. Activity

A radioactive material, or source, consists of a very large number of atoms, which do not all decay at the same time. For every radionuclide, there is a probability that disintegration will occur, known as that radionuclide's **decay constant**. This is noted, as a rule, by the Greek letter λ. This quantity stands for the probability that one nucleus will disintegrate, for every second that passes. Thus, so long as a nucleus has not undergone disintegration, the probability λ that this nucleus will do so during the next second remains the same. The number of disintegrations that occur, per unit time, is the product of the number of radioactive nuclei present, N, by the decay constant, λ; this number of disintegrations per unit time is the **activity** of that radioactive source:

$$A = \lambda N$$

Its activity A serves to characterize a particular source, and is a "measure" of its radioactivity.

In the International System of Units (SI), the unit of activity is the becquerel (symbol: Bq):

$$1 \text{ becquerel} = 1 \text{ disintegration per second}$$

The becquerel is an altogether tiny unit, and, in practice, multiples of the becquerel are generally used:

- the kilobecquerel (kBq): 1 kBq = 10^3 Bq;
- the megabecquerel (MBq): 1 MBq = 10^6 Bq;
- the gigabecquerel (GBq): 1 GBq = 10^9 Bq;
- the terabecquerel (TBq): 1 TBq = 10^{12} Bq.

The earlier unit initially adopted, and still fairly widely used, though it remains outside the SI system of units, is the curie (Ci). The two units of activity stand in the following relation:

$$1 \text{ Ci} = 3.7 \cdot 10^{10} \text{ Bq} = 37 \text{ GBq}$$

1.7.2. Emission rate

The emission rate, for radiation, is defined as the number n of occurrences of a radiation, of a particular, well defined type and energy, emitted per unit time.

This quantity is equal to the product of activity by that radiation's emission intensity:

$$n = A \times \frac{I}{100}$$

where n is expressed in reciprocal seconds (s^{-1}), and I as a percentage (%).

1.7.3. Radioactive decay and half-life

A radioactive source contains a large number of radioactive nuclei; this number will become smaller, as disintegrations take place. The activity of a radioactive source thus decreases over time.

This decrease, or decay, in activity depends on the radionuclide's decay constant λ, the units for which are units of reciprocal time. The preferred quantity, for the purposes of measuring radioactive decay, is the radionuclide's half-life, as a rule noted $T_{1/2}$, this being a time interval that is more practical to use, and more readily to be found in tables.

The radioactive half-life, for a source, is the time required for the number of radioactive nuclei initially present to decrease by one half. This is a characteristic quantity, for a given radioactive isotope, the value for which does not vary.

Depending on the radionuclide, half-life values may correspond to widely different intervals, ranging from a fraction of a second to billions of years.

Examples: ^{131}I $T_{1/2} = 8$ days

^{238}U $T_{1/2} = 4.5$ billion years

The half-lives for the more commonly encountered radionuclides can be found in many data tables.

The value for the half-life $T_{1/2}$ stands in the following relation to that of the decay constant λ:

$$T_{1/2} = \frac{\ln 2}{\lambda}$$

Thus, the activity of any source may now be calculated in two different fashions:

$$A = \lambda N$$

or:

$$A = \frac{\ln 2}{T_{1/2}} N$$

where $T_{1/2}$ is the half-life, expressed in seconds, and λ is the decay constant, for the radioisotope considered.

This formula shows that activity is directly proportional to the number of radioactive atoms present in the source.

Using the definition for the half-life, it now becomes possible to calculate directly the residual activity after an elapsed time that is equal to an integral number of half-lives.

Setting A_0 to be the initial value of activity:

After 1 half-life, residual activity is equal to: $A_1 = \frac{A_0}{2}$
After 2 half-lives: $A_2 = \frac{A_1}{2} = \frac{A_0}{4} = \frac{A_0}{2^2}$
After 3 half-lives: $A_3 = \frac{A_2}{2} = \frac{A_0}{8} = \frac{A_0}{2^3}$
And after n half-lives: $A_n = \frac{A_0}{2^n}$

This relation shows that, after 7 half-lives, residual activity is down to about one hundredth initial activity (1/128), and, after 10 half-lives, it stands at one thousandth initial activity (1/1024).

If it is desired to determine a source's activity after any, arbitrary elapsed decay time t, not equal to an integral number of half-lives, the general relation for radioactive decay must be used:

$$A = A_0 \times e^{-\frac{t}{T_{1/2}} \times \ln 2} \qquad \text{i.e.:} \; A = A_0 \times e^{-\frac{t}{T_{1/2}} \times 0.693}$$

where A_0 stands for the initial value of activity, t for the elapsed decay time, from the initial point in time ($t = 0$).

When using this formula, care must be taken that t, and $T_{1/2}$ are both expressed in the same units.

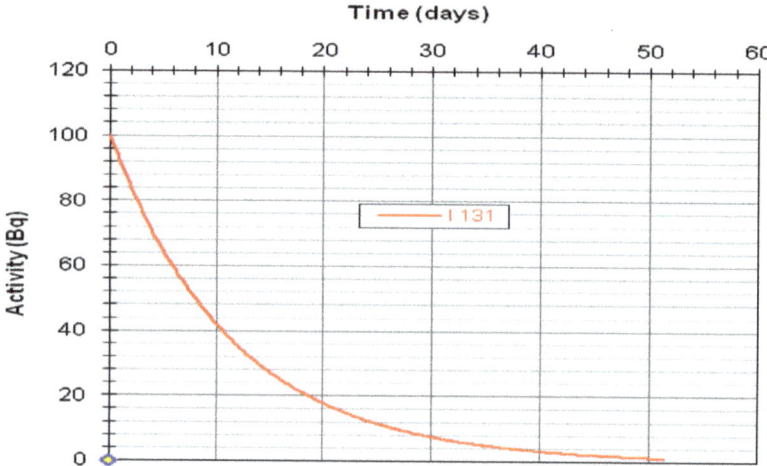

Figure 1.9. Plot of the radioactive decay of iodine-131.

The exponential law set out above, when plotted on semilogarithmic graph paper, yields a straight line, which is straightforwardly drawn, starting at a point $\frac{A}{A_0} = 1$ at the initial time, through to a point $\frac{A}{A_0} = \frac{1}{2^n}$ at time $t = nT_{1/2}$. Figure 1.9 plots the radioactive decay of iodine-131, which has a half-life of 8 days.

Let us apply these relations to the following exercise:

Assume we have a ^{32}P source, with a half-life of 14 days, that has an activity of 1 GBq. We want to calculate its activity after 28 days, and further after 50 days.

After 28 days: this decay time is equal to 2 half-lives. Activity will therefore have decreased to one quarter of its initial value:

$$A = 250 \text{ MBq} \quad (= 0.25 \text{ GBq})$$

After 50 days: this decay time is not equal to an integral number of half-lives. We therefore need to use the general formula for radioactive decay:

$$A = A_0 \times e^{-\frac{t}{T_{1/2}} \times 0.693}$$

which gives the result:

$$A = 10^9 \times e^{-\frac{50}{14} \times 0.693} = 84 \text{ MBq} (= 0.084 \text{ GBq})$$

1.7.4. Radioactive series

1.7.4.1. Two-nuclide decay series

A "radioactive series" is said to arise when the daughter nucleus is also radioactive.
For the purposes of considering what evolution the second nuclide undergoes, this is referred to as a two-nuclide decay series, or decay chain.

1 – Radioactivity and nuclear physics

Let us examine such two-nuclide decay series, by looking at the material balance set out below:

$$1 \xrightarrow{T_1} 2 \xrightarrow{T_2} 3$$

In this case, the activity of nuclide 2 increases, even as that of nuclide 1 decreases, however, since nuclide 2 is itself radioactive, its activity depends both on the half-life of nuclide 1, and its own half-life.

Assuming that, at $t = 0$, only nuclide 1 is present, with its own initial activity (noted $A_{1,0}$), the activity of nuclide 2 (noted A_2), at any time, may be calculated by means of the following general relation:

$$A_2 = \frac{T_1}{T_1 - T_2} A_{1,0} \times \left(e^{-\frac{t}{T_1} \ln 2} - e^{-\frac{t}{T_2} \ln 2} \right)$$

The mathematical analysis of this relation, yielding as it does A_2 as a function of time, shows that A_2 increases, goes through a maximum value, and then decreases back again. It can be shown that, at the point in time when A_2 reaches its maximum value, $A_2 = A_1$.

If T_1 is longer than T_2

Specifically, in cases where T_1 is longer than T_2, after the point when A_2 has reached its maximum value, nuclides 1 and 2 stand in equilibrium, and are said to be in **transient equilibrium**: the ratio of activities, A_1/A_2, remains constant, and A_2 varies thereafter with an apparent half-life equal to that of nuclide 1 (T_1). The general relation may then be simplified, to the following form:

$$A_2 = \frac{T_1}{T_1 - T_2} A_{1,0} e^{-\frac{t}{T_1} \ln 2}$$

It will be seen that, aside from the factor $\frac{T_1}{T_1 - T_2}$, the daughter's activity subsequently varies in step with the decay of the parent nuclide.

Example:

$$^{140}\text{Ba} \xrightarrow{T_1 = 13\,d} {}^{140}\text{La} \xrightarrow{T_2 = 1.68\,d} {}^{140}\text{Ce}$$

Figure 1.10 plots the variations in relative activity for barium-140 and lanthanum-140, as a function of time.

In practice, a decay series of this type may be used to "produce" a short-lived radionuclide, and to make it readily available by means of chemical separation (elution across an ion-exchange column).

The series decay of molybdenum-99, in particular, is very widely used in nuclear medicine, to ensure supplies of the daughter radionuclide, technetium-99m:

$$^{99}\text{Mo} \longrightarrow {}^{99m}\text{Tc} \longrightarrow {}^{99}\text{Tc}$$

Molybdenum-99 has a half-life of 66 hours, technetium-99m a half-life of 6 hours. The maximum activity for technetium-99m is achieved after 23 hours or so. From what has been said above, so long as molybdenum-99 and technetium-99m are not separated, technetium-99m then varies at a rate in accordance with the half-life of molybdenum-99, viz. 66 hours. This ensures a much more prolonged availability of the product to end-users.

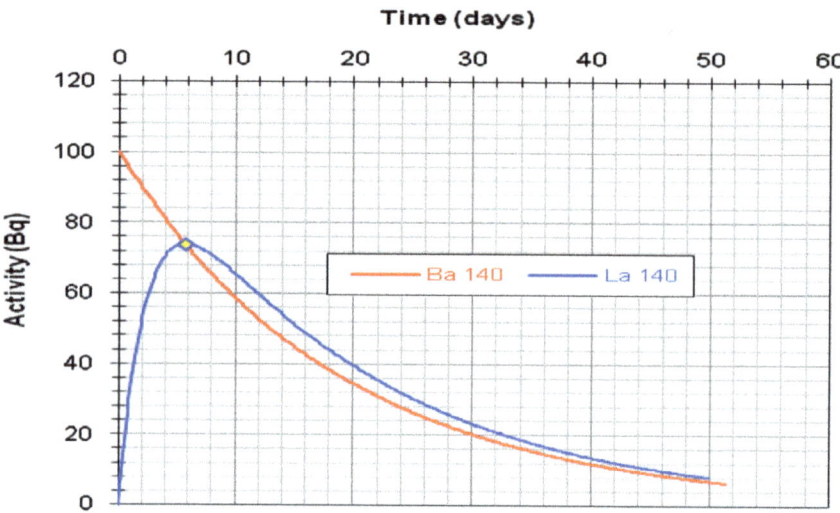

Figure 1.10. Variations in relative activity for barium-140 and lanthanum-140.

In cases where T_1 is much longer than T_2, the ratio $\frac{T_1}{T_1-T_2}$ may be taken as being close to 1. The daughter's activity, A_2, effectively varies, ultimately, with the same value as the parent's activity: this is known as "secular" equilibrium:

$$A_2 = A_{1,0} e^{-\frac{t}{T_1} \ln 2}$$

If T_2 is longer than T_1

Specifically in cases where T_2 is longer than T_1, after the point when A_2 has reached its maximum value, nuclide 1 swiftly decays away, yielding nuclide 2. Thus, no transient equilibrium is set up, between nuclide 1, and nuclide 2. The activity of nuclide 2 thereafter varies in accordance with its own half-life. The general relation may then be simplified as follows:

$$A_2 = \frac{T_1}{T_2 - T_1} A_{1,0} e^{-\frac{t}{T_2} \ln 2}$$

It will be seen that, aside from the factor $\frac{T_1}{T_2-T_1}$, the daughter's activity subsequently varies in accordance with its own decay law.

Example:

$$^{131}_{52}\text{Te} \xrightarrow{T_1=1.25\text{ d}} {}^{131}_{53}\text{I} \xrightarrow{T_2=8\text{ d}} {}^{131}_{54}\text{Xe}$$

Figure 1.11 plots the variations in relative activity for tellurium-131 and iodine-131, as a function of time.

1.7.4.2. The n-nuclide decay series

The evolution of a particular nuclide may also be investigated when a number of disintegrations are involved, occurring as part of a single decay series. The nuclide of interest

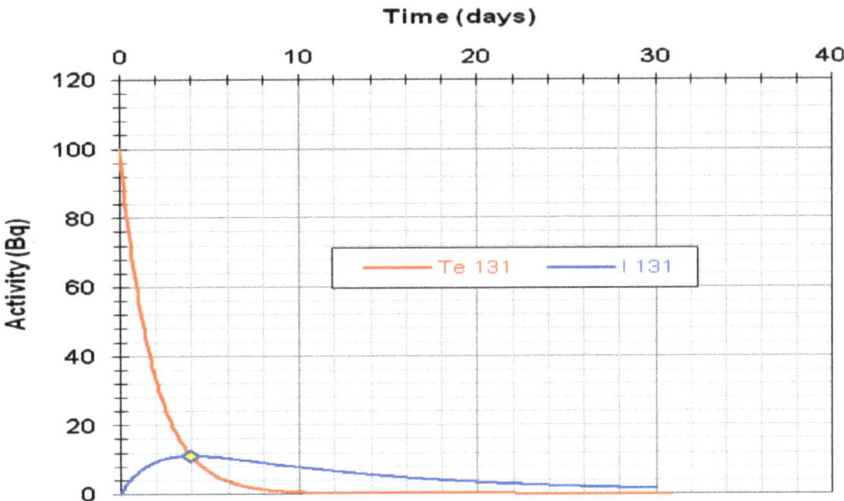

Figure 1.11. Variations in relative activity for tellurium-131 and iodine-131.

may arise as the third in the series, or even further down the line. For this case, a general formula is available, that may be applied to the *n*-nuclide decay series. This equation, or rather this set of equations – known as the "Bateman equations," after the mathematician who originally studied these formulas for the *n*-nuclide case – makes it possible to calculate the activity of any nuclide in the chain, through to the *n*th nuclide, at any time *t*. The general relation takes the following form:

$$A_n = N_{1,0}\lambda_1 \ldots \lambda_n \left[\frac{1}{(\lambda_2 - \lambda_1)} \cdots \frac{1}{(\lambda_n - \lambda_1)} e^{-\lambda_1 t} + \cdots + \frac{1}{(\lambda_1 - \lambda_n)} \cdots \frac{1}{(\lambda_{n-1} - \lambda_n)} e^{-\lambda_n t} \right]$$

Remember that, for practical purposes, λ may always be substituted for by $\frac{\ln 2}{T_{1/2}}$.

It will be noted that the number of exponential terms correspond to the number of nuclides considered, and that the coefficients assigned to each exponential term have very similar denominators.

If one seeks to calculate the activity of the third nuclide in the series, noted A_3, assuming that, at the initial time ($t = 0$), only nuclide 1 is present, the above equation becomes:

$$A_3 = N_{1,0}\lambda_1\lambda_2\lambda_3 \left[\frac{1}{(\lambda_2 - \lambda_1)} \frac{1}{(\lambda_3 - \lambda_1)} e^{-\lambda_1 t} + \frac{1}{(\lambda_1 - \lambda_2)} \frac{1}{(\lambda_3 - \lambda_2)} e^{-\lambda_2 t} \right. $$
$$\left. + \frac{1}{(\lambda_1 - \lambda_3)} \frac{1}{(\lambda_2 - \lambda_3)} e^{-\lambda_3 t} \right]$$

Fortunately, in most *n*-nuclide problems, the position can usually be simplified, by considering the longest radioactive half-lives, which, in many cases, allows the equation to be brought back to the two-nuclide situation.

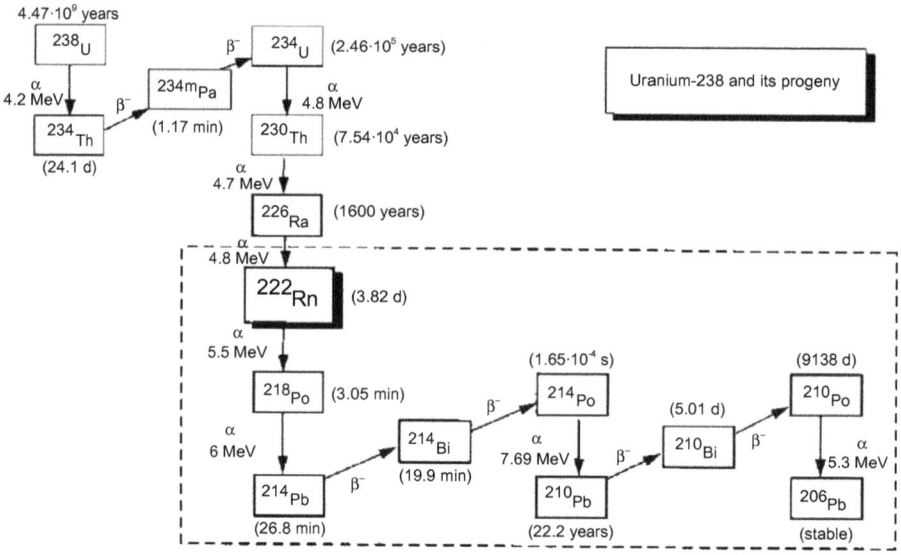

Figure 1.12. An instance of radioactive series: the uranium-238 series.

There are three naturally occurring families forming n-nuclide radioactive series, yielded by radioactive progenitor elements that were present when the Earth was formed, such as e.g. the uranium-238 series, shown in Figure 1.12.

The dashed-line box encloses the progeny of radon-222, a gas that is liable to escape into the atmosphere (see Chapter 6, "Uses of sources of ionising radiation").

1.7.5. Activity–mass relationship

Consider a source consisting of radioactive atoms, of a type $^{\mathcal{A}}X$, with a half-life $T_{1/2}$. By definition, the mass of N_A atoms has a value that differs but little from $\mathcal{A}$ grams. The mass of N atoms of $^{\mathcal{A}}X$ may thus be taken to stand equal to:

$$m = \mathcal{A}\frac{N}{N_A}$$

where m is expressed in grams, and $N_A = 6.02 \cdot 10^{23}$ (Avogadro's number).

And, by carrying over this result into the relation derived in section 7.3:

$$A = N_A \frac{m}{\mathcal{A}} \frac{\ln 2}{T_{1/2}}$$

where A is the activity of a mass m of nuclide $^{\mathcal{A}}X$, of mass number $\mathcal{A}$. (In this section, for the purposes of this formula, we are using a distinct typeface to denote activity A, to distinguish it from mass number $\mathcal{A}$.)

The above formula shows that, for a given level of activity, the mass of any particular source is directly proportional to its half-life. Setting an activity level A equal to $3.7 \cdot 10^{10}$ Bq, this relation, when applied to the following radionuclides, yields the following values for *m*:

$$^{238}\text{U} \quad T_{1/2} = 4.5 \cdot 10^9 \text{ years} \quad m = 3000 \text{ kg}$$

$$^{226}\text{Ra} \quad T_{1/2} = 1620 \text{ years} \quad m = 10^{-3} \text{ kg}$$

$$^{32}\text{P} \quad T_{1/2} = 14 \text{ days} \quad m = 3.4 \cdot 10^{-9} \text{ kg}$$

$$^{99m}\text{Tc} \quad T_{1/2} = 6 \text{ hours} \quad m = 0.19 \cdot 10^{-9} \text{ kg}$$

1.7.6. Production of radionuclides

Short-lived radionuclides must, as a rule, be specially produced, in order to meet the needs of industry, or medicine (see Chapter 6, "Uses of sources of ionising radiation"). They are consequently referred to as "artificial radionuclides." Many of these isotopes are generated close to a nuclear reactor core, or in the vicinity of accelerators, or cyclotrons, by making use of nuclear reactions, involving neutron activation of, or charged-particle interactions with, stable nuclei.

A nuclear reaction takes the form – barring some few exceptional cases:

$$^{A_1}_{Z_1}a + ^{A}_{Z}X \longrightarrow ^{A_2}_{Z_2}b + ^{A'}_{Z'}Y$$

which reaction may also be written out in the following manner: $^{A}_{Z}X \left(^{A_1}_{Z_1}a, ^{A_2}_{Z_2}b \right) ^{A'}_{Z'}Y$

where a is the particle acting as a projectile (e.g. a neutron, a proton, a deuteron – i.e. a deuterium nucleus – an alpha particle, an electron, or indeed any light particle), X the target nucleus (in many cases that of a stable nuclide, more rarely a radioactive nuclide), b the product particle, Y the product nucleus, which is usually radioactive.

The nuclear upheaval brought about by a nuclear reaction is virtually instantaneous. Experiments show that every nuclear reaction complies with three conservation laws:

- conservation of total mass number: $A_1 + A = A_2 + A'$;

- conservation of total atomic number: $Z_1 + Z = Z_2 + Z'$;

- conservation of total energy: the sum total of kinetic energy, and mass-energy before transformation is equal to the sum total of kinetic energy, and mass-energy after transformation, for the components involved.

1.7.6.1. Nuclear reactions: neutron activation

This section describes the (n,γ) nuclear reaction, also known as **neutron activation**. It should be borne in mind that some research nuclear reactors are specifically designed to produce artificial radioisotopes, by neutron activation. This is the technique used, in particular, to produce cobalt-60 from stable cobalt-59.

An (n,γ) nuclear reaction may be represented, as a rule, in terms of the following general relation:

$$X + n \rightarrow Y + \gamma$$

The number of Y atoms generated through irradiation by a neutron beam, per unit time dt, noted dN_Y^+, is proportional to the number of atoms in the target (N_X), to the incident neutron flux Φ_n (i.e. the number of incoming neutrons per unit surface, per unit time), and to the probability for the reaction σ_{XY} (known as the cross-section for that reaction, expressed in square centimeters [cm^2], or in barns [b]: 1 b = 10^{-24} cm^2):

$$dN_Y^+ = N_X \cdot \Phi_n \cdot \sigma_{XY} \cdot dt$$

During the same time interval dt, since the nuclide Y that is being generated is radioactive, the number of Y atoms that disappear through radioactive decay, noted dN_Y^-, is equal to:

$$dN_Y^- = N_Y \cdot \lambda \cdot dt$$

where λ is the decay constant for the radionuclide Y being generated.

The net number of Y atoms actually obtained, over a time interval dt, thus stands equal to:

$$dN_Y = dN_Y^+ - dN_Y^-$$

$$dN_Y = N_X \cdot \Phi_n \cdot \sigma_{XY} \cdot dt - N_Y \cdot \lambda \cdot dt$$

or: $\quad dN_Y/dt + N_Y \cdot \lambda = N_X \cdot \Phi_n \cdot \sigma_{XY}$

Owing to the low values found for the probability for the reaction, N_X may be taken to remain constant during irradiation; so, performing the integration over time, for this equation, yields – assuming that the target is not radioactive at the initial time – the following formula:

$$N_Y(t) = \frac{N_X \cdot \Phi_n \cdot \sigma_{XY}}{\lambda} \times \left(1 - e^{-\lambda t}\right)$$

The number of Y atoms thus generated corresponds to an activity equal to:

$$A = \lambda \cdot N_Y$$

therefore: $\quad A(t) = N_X \cdot \Phi_n \cdot \sigma_{XY} \times (1 - e^{-\lambda t})$

where: A = activity of the radioactive nuclide generated (Bq);
σ_{XY} = cross-section for the reaction (cm^2);
Φ_n = neutron beam flux (cm$^{-2} \cdot$s^{-1});
t = activation time (s);
λ = decay constant for nuclide Y. It will be recalled that $\lambda = \ln 2/T_{1/2}$, where $T_{1/2}$ is the half-life.

If the activation time is long, compared to the half-life of the radioactive nuclide generated, the exponential term tends to zero. For t longer than $10 T_{1/2}$, the formula may thus be simplified to the following:

$$A(t) = N_X \cdot \Phi_n \cdot \sigma_{XY}$$

Example:

Consider a 30-mg cobalt-59 target. The cross-section for the ^{59}Co(n,γ)^{60}Co reaction is found to be equal to $20 \cdot 10^{-24}$ cm^2, while cobalt-60 has a half-life of 5.27 years. If a neutron flux Φ_n of $5 \cdot 10^{12}$ neutrons·cm^{-2}·s^{-1} is used, what is the level of cobalt-60 activity that will be available after an irradiation time of 30 days?

Let us refer back to the formula:

$$A(t) = N_X \cdot \Phi_n \cdot \sigma_{XY} \times (1 - e^{-\lambda t})$$

Bearing in mind that 59 grams cobalt-59 contain one mole of atoms, i.e. $6.02 \cdot 10^{23}$ atoms, it will be found that 30 mg cobalt-59 contain $3.06 \cdot 10^{20}$ atoms. Numerical calculation then yields:

$$A = 3.06 \cdot 10^{20} \times 20 \cdot 10^{-24} \times 5 \cdot 10^{12} \times \left(1 - e^{-\frac{30}{5.27 \times 365} \times \ln 2}\right) = 329 \cdot 10^6 \text{ Bq}$$

1.7.6.2. Nuclear reactions: charged-particle reactions

Charged particles also serve as projectiles, allowing artificial radionuclides to be produced. Light nuclei are used, raised to varyingly high levels of energy by means of accelerators, to strike the target nuclei. If the particle used is positively charged (proton, deuteron, alpha particle…), the energy imparted to it must be sufficiently high to enable it to overcome the Coulomb repulsive force that arises between it, and the target nucleus. This force becomes all the stronger, the higher the electric charges borne by the incident particle, and the target nucleus.

If excessively high energy levels are employed, the nuclear reactions may prove altogether too violent, resulting in a huge variety of new particles being generated, and not allowing the desired radionuclide to be obtained. Production of radionuclides is, in virtually all cases, carried out in low-energy reaction conditions, yielding straightforward rearrangements of the protons and neutrons in the target nucleus.

An accelerator is not always required to obtain charged-particle nuclear reactions. By way of example, we can look at a nuclear reaction that makes it possible to set up a neutron source, by mixing an alpha-emitter americium source with an isotope of low atomic weight, e.g. beryllium:

$$^4_2\alpha + ^9_4\text{Be} \rightarrow ^0_1\text{n} + ^{12}_6\text{C}$$

or:

$$^9_4\text{Be}(^4_2\alpha, ^0_1\text{n})^{12}_6\text{C}$$

The Be–Am mix takes the form of a powdered blend. In this case, the yield is low, as production of some 30 neutrons may be anticipated, for every million alpha particles emitted.

Many other nuclear reactions may be used, the more so since applications of radioactivity are as numerous as they are diverse. It would be altogether unfeasible to mention all of these applications, together with the nuclear reactions involved. It should be pointed out, however, that, in the field of medicine, widespread benefits have accrued from use of the properties of such radiation, for therapeutic, or diagnostic purposes (see Chapter 6, "Uses of sources of ionising radiation").

1.7.6.3. Nuclear reactions: fission

Fission reactions are processes whereby a heavy nucleus breaks up into two lighter nuclei, known as "fission products." Two of the more widely know instances of heavy nuclei liable to undergo fission are uranium-235, and plutonium-239. This reaction occurs when a neutron collides with the nucleus. This phenomenon of **induced fission** is quite well known, as it provides the basis for the operation of nuclear reactors.

By way of example, the fission reaction, by neutron absorption, of uranium-235 may be written as follows:

$$^{0}_{1}n + ^{235}_{92}U \rightarrow ^{A_1}_{Z_1}X + ^{A_2}_{Z_2}Y + k^{0}_{1}n$$

This fission reaction yields the emission of 2 or 3 neutrons (i.e. $k = 2$ or 3), and an energy release of around 200 MeV, nearly entirely carried off by the fission products, noted X and Y. The fission products are radioactive as a rule. Heavy nuclei that are liable to undergo fission in this manner are said to be "fissile" they invariably involve an atomic number higher than, or equal to, 89 (actinium): they form a family, known as the actinide series. These radionuclides all exhibit chemical properties close to those of actinium.

Further, very heavy nuclei belonging to the actinide family, featuring as a rule mass numbers higher than 230, i.e. isotopes of thorium and heavier elements, may undergo **spontaneous fission**. Such fission phenomena are spontaneous, in that they require no extraneous input. The nuclei involved do not require the extra energy contributed by a neutron, or indeed any other particle, to undergo fission. The outcome, as for induced fission, is the breakup of a nucleus into two lighter fragments. This type of fission stands as a process competing with alpha decay, since it involves nuclei that seek to reduce their excess mass. In most cases, alpha decay is markedly predominant, compared to spontaneous fission. Thus, californium-252 undergoes spontaneous fission in just 3.1% of cases, whereas alpha decay accounts for 96.9% of ^{252}Cf decay events. Some few exceptions do stand out, however, e.g. californium-254, for which spontaneous fission accounts for 99.7% of decays.

1.8. Check your knowledge

You may now test what you have learned in studying this chapter, by finding the answers to the following questions:

1. **In the following list of nuclides, which ones are isotopes?**

    $^{197}_{79}Au \quad ^{197}_{80}Hg \quad ^{194}_{78}Pt \quad ^{198}_{79}Au \quad ^{192}_{77}Ir \quad ^{197}_{81}Tl \quad ^{200}_{79}Au$

    Answer:

    The isotopes of gold (Au).

2. **What types of ionising radiation are emitted following electron-capture decay? Very briefly, account for their origin.**

    Answer:

    No detectable radiation is emitted by the nucleus, subsequent to electron-capture decay. The only ionising radiations yielded by this event are X-rays, together with

Auger electrons, resulting from the rearrangement of the electron cloud, due to the disappearance of a strongly bound electron.

3. **How would you characterize the energy spectrum for the particles emitted in beta decay? What is the value that is shown in tables, to characterize the energy of emitted beta particles?**

   Answer:

   This is a continuous energy spectrum, ranging from 0 to $E_{\beta\,max}$. Tables always indicate $E_{\beta\,max}$.

4. **How would you characterize the energy spectrum for the particles emitted in alpha decay? Does gamma decay yield a similar spectrum?**

   Answer:

   This is a line spectrum, with characteristic lines corresponding to particle energies. Gamma decay yields the same type of spectrum.

5. **Consider a cobalt-56 radioactive source, with an activity of 50 kBq. It decays by beta-plus emission, and electron capture. It further emits gamma rays. For the various radiations of highest intensity, values for the characteristic quantities are as follows:**

   $E_{\beta^+ max} = 1.46$ MeV, $I_{\beta^+} = 18\%$

   $E_{\gamma_1} = 846$ keV, $I_{\gamma_1} = 100\%$

   $E_{\gamma_2} = 1238$ keV, $I_{\gamma_2} = 67\%$

   $E_{\gamma_3} = 2.6$ MeV $I_{\gamma_3} = 17\%$

   1) What is the emission intensity for electron-capture decay?

   2) What are the emission rates for the beta-plus radiation, and for γ_2 rays?

   Answers:

   1) $I_{EC} = 100 - I_{\beta^+} = 100 - 18 = 82\%$.

   2) $n_{\beta^+} = A \times \frac{I_{\beta^+}}{100} = 50 \cdot 10^3 \times 0.18 = 9000\ \beta^+ \cdot s^{-1}$.

   $n_{\gamma_2} = A \times \frac{I_{\gamma_2}}{100} = 50 \cdot 10^3 \times 0.67 = 33\,500\ \gamma_2 \cdot s^{-1}$.

6. **Assume you have an antimony-124 radioactive source, with an initial activity of $2.5 \cdot 10^5$ Bq. Its half-life being $T_{1/2} = 60$ days:**

   1) What will the source's activity be after 4 months?

   2) What will the source's activity be after 321 days?

   Answers:

   1) We may note that 4 months, i.e. 120 days, correspond to 2 half-lives (viz. 2 × 60 days); therefore:

   $$A_{4months} = \frac{A_0}{2^2} = \frac{2.5 \cdot 10^5}{4} = 62500\ Bq.$$

2) $A_{321\,d} = A_0 \times e^{-\frac{t}{T_{1/2}} \times \ln 2} = 2.5 \cdot 10^5 \times e^{-\frac{321}{60} \times \ln 2} = 6130$ Bq.

7. **Consider a phosphorus-33 radioactive source, the half-life for which is 25 days. If the source has an initial activity of 40·10³ kBq, how long will it take for its activity to come down to 500 kBq?**

Answer:

As we have seen: $A = A_0 \times e^{-\frac{t}{T_{1/2}} \times \ln 2}$.

We may therefore write: $t = \frac{\ln\left(\frac{A_0}{A}\right)}{\ln 2} T_{1/2} = \frac{\ln\left(\frac{40 \cdot 10^5}{500}\right)}{\ln 2} \times 25 = 158$ days.

8. **Copper-64 decays by beta-minus emission, beta-plus emission, and electron capture. The characteristic quantities, for the various types of radiation emitted, have the following values:**

$E_{\beta^-\text{max}} = 578$ keV, $I_{\beta^-} = 36.8\%$

$E_{\beta^+\text{max}} = 653$ keV, $I_{\beta^+} = 18.1\%$

$E_\gamma = 1345$ keV, $I_\gamma = 0.5\%$

1) What is the emission intensity for electron capture?

2) What is the emission rate for the 1345-keV gamma ray, in a source with an activity of 7.4 MBq?

3) If that same source is enclosed by shielding which fully absorbs beta-plus and beta-minus radiation, what is the emission rate for the 511-keV annihilation radiation? (*It is advisable to have read Chapter 2, "Interaction of radiation with matter" before answering this question.*)

4) Since copper-64 has a half-life of 12.7 hours, what will the source's activity be after 38.1 hours?

Answers:

1) $I_{EC} = 100 - I_{\beta^-} - I_{\beta^+} = 100 - 36.8 - 18.1 = 45.1\%$.

2) $n_\gamma = A \times \frac{I_\gamma}{100} = 7.4 \cdot 10^6 \times \frac{0.5}{100} = 3.7 \cdot 10^4 \ \gamma \cdot s^{-1}$.

3) There are 2 × 511-keV annihilation photons for every beta-plus particle emitted; therefore:

$$n_{511\text{-keV photon}} = 2A \times \frac{I_{\beta^+}}{100} = 2 \times 7.4 \cdot 10^6 \times \frac{18.1}{100} = 2.68 \cdot 10^6 \text{ photons} \cdot s^{-1}.$$

4) We may note that 38.1 hours correspond to 3 half-lives; therefore:

$$A_{38.1h} = \frac{A_0}{2^3} = \frac{7.4}{8} = 0.93 \text{ MBq}.$$

1 – Radioactivity and nuclear physics

9. Consider a source of 1 gram radium-226.

1) Calculate the activity of this source. What do you find?
2) To what mass of tritium would this activity correspond?

Data: $N_A = 6.02 \cdot 10^{23}$ atoms

$T_{1/2}$ (radium-226) = 1582.89 years

$T_{1/2}$ (tritium) = 12.34 years

Answers:
1) $A = N_A \frac{m}{A} \frac{\ln 2}{T_{1/2}} = 6.02 \cdot 10^{23} \times \frac{1}{226} \times \frac{\ln 2}{1582.89 \times 365.25 \times 24 \times 3600} = 3.7 \cdot 10^{10}$ Bq.
This activity is equal to 1 curie. The curie thus corresponds to the activity of 1 gram radium-226.
2) As we have seen: $A = N_A \frac{m}{A} \frac{\ln 2}{T_{1/2}}$; we may therefore write:

$$m = \frac{A T_{1/2}}{\ln 2} \frac{A}{N_A} = \frac{3.7 \cdot 10^{10} \times 12.34 \times 365.25 \times 24 \times 3600}{\ln 2} \times \frac{3}{6.02 \cdot 10^{23}} = 10^{-4} \text{g} = 0.1 \text{ mg}.$$

Annex 1: Periodic table of the elements (source: IUPAC).

Interaction of ionising radiation with matter

Philippe Massiot, Hugues Bruchet

Introduction

The preceding chapter set out to define the phenomenon of radioactivity, and described the various processes whereby radioactive transformations take place. Thus, we saw that excess energy residing in the nuclei of radioactive isotopes is released in the form of ionising radiation of various kinds, endowed with a range of energies, and diverse in nature. The nature of this radiation depends both on the amount of excess energy present in the unstable nucleus, and equally on the structure exhibited by that nucleus (i.e. depending on whether it is a "heavy" nucleus, involving a high atomic number Z, or a proton-rich, or neutron-rich nucleus).

Whereas that Chapter 1, "Radioactivity and nuclear physics", had restricted its scope, in spatial terms, to consideration of the radioactive source, the purpose of the present chapter is to provide a description of the behavior of ionising radiation once it has arisen, subsequent to nuclear decay, whether by way of a disintegration process, or deexcitation of the radioactive nucleus.

Such radiation, endowed as it is with energy, goes on to interact with the basic structural constituents of matter, i.e. essentially with electrons, and atomic nuclei. In this interaction, radiation transfers all or part of its energy to matter. Conversely, any matter involved in this process undergoes alterations, due to that interaction.

It should be emphasized, at this point, that a wide range of "radiation–matter" interactions can and do occur, depending on the nature of the radiation involved, and the amount of energy it carries.

We will begin by setting out the definition for, and classification of, ionising radiation, prior to reviewing the interaction of charged particles with matter – this will entail a distinction between light charged particles (chiefly electrons), and heavy charged particles (mainly alpha particles). This section will provide tools that allow the ranges of such particles inside matter to be calculated.

The following section will go on to consider the case of electromagnetic waves – i.e. electromagnetic radiation. A law is found to exist, that serves to describe the attenuation of such radiation in matter.

The chapter will then round off by describing the interaction of neutrons with matter.

Radiation, of any kind, may only be detected, and characterized, by way of its interactions with matter, viz. the materials through which it propagates. This chapter thus stands as an essential preliminary for Chapter 5, "Detection and measurement of ionising radiation".

2.1. Ionising radiation: definition and classification

The term "radiation" covers any process whereby energy is emitted, or transmitted, in the form of electromagnetic waves, or particles.

Radiation is referred to as "ionising radiation" when it has the ability to strip electrons from matter. A more comprehensive definition may be given, as follows: ionising radiation is the transport of energy in the form of particles, or electromagnetic waves of wavelengths shorter than or equal to 100 nanometers – i.e. of frequencies ranging down to $3 \cdot 10^{15}$ hertz – found to have the ability to produce ions, whether directly, or indirectly.

This definition, however unwieldy it may seem, at first blush, does point us to two crucial features:

- first, the fact that ionising radiation consists in the transport of **energy**: in the absence of energy, no ionising radiation can occur – then there is no call for radiation protection, since it is the transfer of radiation energy to it that causes damage to arise in matter;

- second, the **nature** of that radiation: this is quite crucial, since the types of interaction liable to arise with matter depend on it. To put it another way: a particle, or an electromagnetic wave yields its energy in a manner that is altogether characteristic of it.

A classification of radiations may be set out as follows, according to their nature, and the energies involved:

Two kinds of radiation are thus seen to arise:

- **particles**, or **particulate radiation**, which involve a rest mass, exhibited by the particle. The total energy of such particles is given by Einstein's formula:

$$E = mc^2$$

where m is the particle's mass, c the velocity ("celerity") of light ($3 \cdot 10^8$ m·s^{-1}).

This total energy may in turn be broken down into two terms, as follows:

$$E = mc^2 = T + m_0 c^2$$

where T is the particle's kinetic energy, $m_0 c^2$ the energy equivalent of that same particle's rest mass;

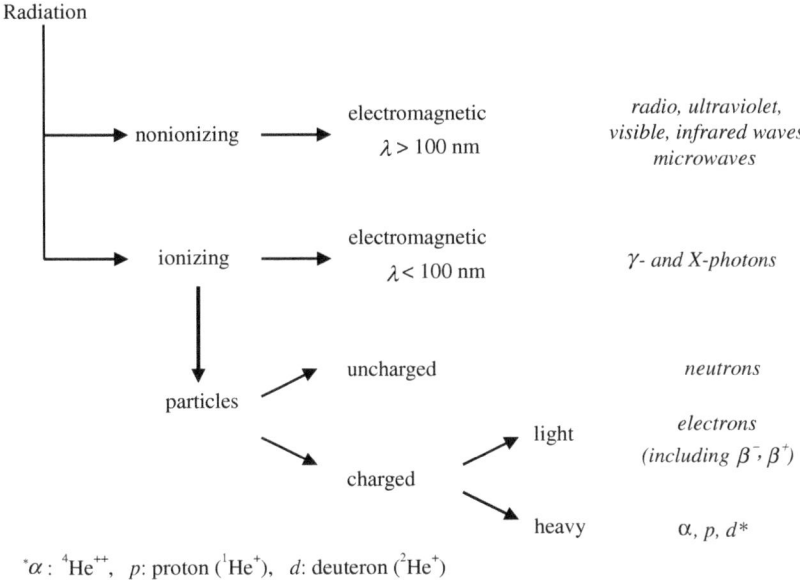

*α: ^{4}He^{++}, p: proton (^{1}He$^+$), d: deuteron (^{2}He$^+$)

Figure 2.1. Classification of radiations.

- **electromagnetic radiation**, consisting of a flow of photons, which therefore involves no rest mass. The expression "propagating energy" is sometimes used to refer to such radiation. Its energy is given by the formula:

$$E = h\nu = h\frac{c}{\lambda} \qquad (2.1)$$

where h is Planck's constant (6.6·10^{-34} J·s), ν the radiation's frequency (in s^{-1}), equal to the ratio of c divided by the radiation's wavelength λ (in m).

In most instances, in what follows, electromagnetic radiation will be characterized either in terms of its energy E, or of its wavelength (as in Figure 2.1).

By definition, nonionising radiation is radiation endowed with insufficient energy to ionize matter. From Figure 2.1, it will be seen that this type of radiation consists, as a whole, of electromagnetic radiations of wavelengths longer than 100 nm.

In contrast, ionising radiation involves a wavelength shorter than 100 nm (this is indeed the value specified in the definition set out above), if it is an electromagnetic radiation. X-rays and γ-rays both come under this category. Using Formula (2.1), the energy equivalent for a wavelength of 100 nm is found to stand equal to 12.4 eV (see the box "Further information", below).

Ionising radiation may equally consist of particles. In that case, a distinction is made between uncharged particles, e.g. neutrons, and charged particles. As regards the latter, mass stands as a crucial criterion, for the purposes of describing the type of interaction that arises: "light" (low-mass) charged particles – chiefly comprising electrons – and heavy charged particles – e.g. α particles, protons, or deuterons (i.e. deuterium nuclei) – will thus be considered separately.

> **Further information**
>
> To work out the value, in electron volts, of the minimum energy that is liable to cause ionisation in matter, all that is required is to apply Formula (2.1), as the value of the corresponding wavelength is known (100 nm). The numerical calculation goes as follows:
>
> $$E = h\frac{c}{\lambda} = 6.6 \cdot 10^{-34} \times \frac{3 \cdot 10^8}{100 \cdot 10^{-9}} = 1.98 \cdot 10^{-18} \text{ J}$$
>
> i.e., in electron volts:
>
> $$E = \frac{1.98 \cdot 10^{-18} \text{ J}}{1.6 \cdot 10^{-19}} = 12.4 \text{ eV}$$

2.2. Interaction of charged particles with matter

2.2.1. General considerations

With respect to matter, the interaction properties exhibited by charged particles chiefly stem from the electric charge they bear. When such a particle enters a medium, it necessarily interacts with all the constituent electrons in the atoms it passes near to, along its path. With regard to this, charged particles are referred to as **directly ionising particles**.

Every interaction of such a particle with an electron involves **a partial transfer** of energy δE, from the incident particle to the electron. Such transfers involve varying quantities of energy, from one interaction to the next – quantities that, as a whole, are truly tiny, compared to the particle's total energy: from a few electron volts (eV) to several kiloelectron volts (keV), however the number of collisions involved is huge.

These energy transfers result in two distinct effects:

- a transfer of energy, imparted to matter in the form of ionisation, or excitation* of atoms lying along the charged particle's path, the greater part of this energy being absorbed locally;

- a gradual, continuous loss of energy, or braking effect, undergone by the particle all along its path, until it ultimately come to a stop, if the thickness of material is sufficient.

Interactions of the "charged particle–electron" type are referred to as collisions. Two types of collision may occur: **ionisation**, and **excitation**. When the interaction is of the "charged particle–nucleus" type, it is termed **bremsstrahlung** ("braking radiation"). The latter phenomenon only arises for high-energy incident electrons.

Ionisation involves a transfer of energy, from an incident particle (alpha particle, electron, positron) to an atom. The energy imparted is sufficient to strip an electron, and eject it from the atom's electron cloud. The outcome is thus the generation, within the material, of one free electron, and one ionized atom (i.e., an ion pair).

Should the ionisation of an atom occur by way of an inner-shell electron (this being less probable than ionisation on an outer shell), an electron from an outer shell fills the

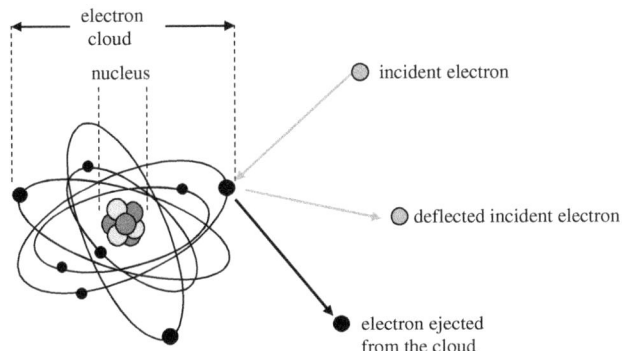

Figure 2.2. Ionisation process.

vacancy thus arising in the inner shell. The energy from this process is released in the form of a photon. The outcome, as a rule, is a cascade of low-energy emissions of electromagnetic radiation, and/or electrons. This is referred to as a rearrangement of the atom's electron cloud.

In some instances, the energy transferred may prove insufficient to strip an electron from the cloud. Another mechanism is then involved, whereby an electron in the cloud may be transferred to a higher energy level, since, in most cases, the **excitation** of an atom occurs by way of an electron from an inner shell.

In that case, the quantity of energy transferred is precisely equal to the difference in binding energies between two electron shells. Thus, no ion pair is generated. The atom is said to become "excited". The electron subsequently reverts to its initial state, energy being released in the form of a light photon, of an energy corresponding to the difference between the two energy levels involved.

2.2.2. *Interaction of electrons with matter*

2.2.2.1. *Ionisation and excitation*

Electrons are low-mass particles, bearing one elementary electric charge, of negative sign for "negatrons", positive for "positrons".

With electrons, **ionisation** and **excitation** are the phenomena most likely to occur. The incident electron thus transfers part of its kinetic energy to an electron bound to an atom; depending on the quantity of energy transferred, either the former, or the latter interaction will occur.

In **ionisation**, the energy yielded by the incident electron is, as a rule, tiny compared to its total kinetic energy. From this, it can be inferred that a great many interactions must occur for the electron to expend its kinetic energy, and "come to rest" in the material.

In the event of ionisation, when the incident electron is a negatron, it becomes indistinguishable, once it has interacted, from the electron ejected from the atom. By convention, the electron endowed with the higher kinetic energy is then identified as the incident electron.

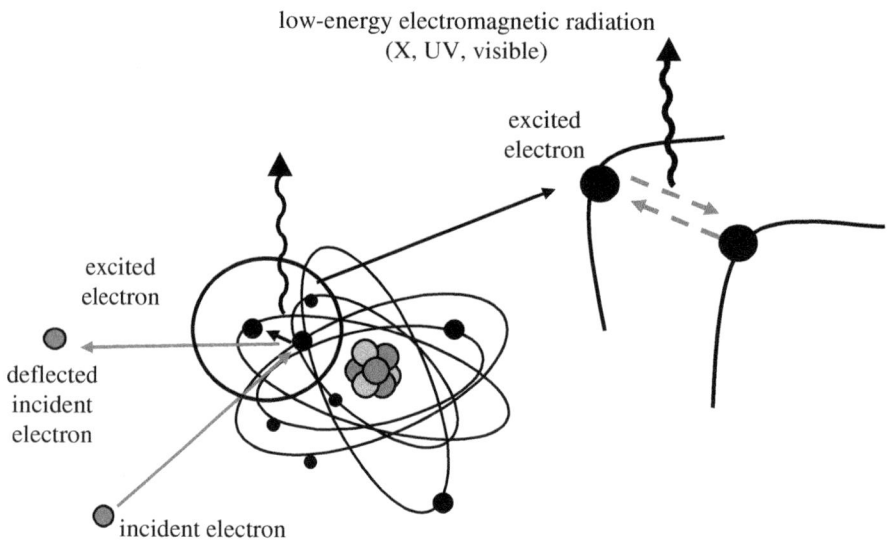

Figure 2.3. Excitation process.

2.2.2.2. Bremsstrahlung

Less frequently, incident electrons are able to interact with atomic nuclei in the medium they travel through. They are then subject to the influence of the nuclei's Coulomb force field: they are thus deflected, yielding part of their energy to the nucleus. This shows up as a slowing down, or braking effect. The resulting energy loss is emitted in the form of X-rays, known as "bremsstrahlung" (a German word meaning "braking radiation").

Bremsstrahlung is a direct consequence of the laws of electromagnetism: any charged particle undergoing acceleration (whether positive, or negative) and deflected in its path emits a quantum of electromagnetic energy, resulting in a loss of energy, and thus a slowing down of the particle.

Thus, incident electrons interacting with atomic **nuclei** in the material they travel through undergo – through the action of the Coulomb field – acceleration, and a path deflection. This shows up, consequently, in the form of bremsstrahlung, which is subsumed under the general phenomenon of X-radiation, being likewise generated by electrons (see Figure 2.4).

This phenomenon becomes significant only for high-energy electrons (viz. of energies higher than 1 MeV) traversing materials consisting of heavy atoms (i.e. of high atomic number Z).

2.2.2.3. An application of bremsstrahlung: the X-ray tube

In order to obtain readily usable X-ray beams, the technological process used involves generating electrons, and directing them onto a target, to yield large amounts of bremsstrahlung.

An electric current passes through a filament, which sends out electrons as it is heated by the current (by thermionic emission, also known as the Edison effect). This filament forms the cathode of the X-ray tube. The electrons thus generated are accelerated by a high

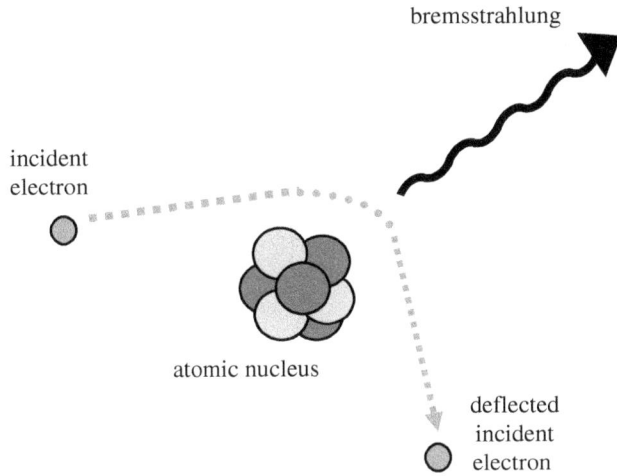

Figure 2.4. Bremsstrahlung.

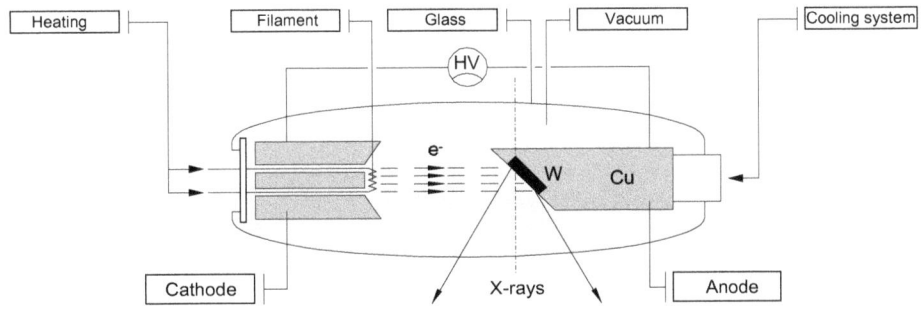

Figure 2.5. Principle of the X-ray tube.

voltage applied across the tube, between the cathode and a target, serving as the anode (this target is sometimes referred to as the "anticathode"), which is raised to a high positive potential, to attract the electrons.

The entire device is enclosed in a sealed glass tube, which is kept under high vacuum, to preclude electron scattering (see Figure 2.5).

To promote bremsstrahlung production, the anode is made of a heat-resistant, high-atomic-number (high-Z) material. Tungsten (atomic number 74, chemical element symbol W) meets admirably both requirements, and is thus widely adopted as the target material in X-ray tubes. Depending on planned applications, the target may equally be made of molybdenum, gold, copper, iron, cobalt…

The bremsstrahlung yield obtained in the target is invariably low (2%). The large number of ionisation and excitation events occasioned by incident electrons, within the target, causes it to heat up heavily. The target is consequently quite commonly embedded in copper, to allow heat diffusion, further calling for an air-, oil-, or water-cooling system.

The anode, in most cases, is set at an angle (slope) relative to the direction of electron travel, and X-rays are then preferentially emitted in a cone, the axis of which varies in orientation according to the anode slope. X-rays pass through the tube wall, exiting via a window made of a light (low-Z) material (beryllium, or aluminum), let into the lead shielding enclosing the tube. This window acts as a barrier for some of the X-rays, particularly the low-energy fraction. Further, in order to restrict the beam to its useful section, a collimator is interposed, after the exit window.

The various materials the beam passes through, as it comes out of the tube, form the tube's "inherent filtration." For the purposes of modifying the energy spectrum of the outgoing beam, further filtration may be interposed, at the tube exit port: this is the added filtration.

2.2.2.4. Linear energy transfer

The three mechanisms outlined in the preceding paragraphs allow electrons to transfer their energy to the material they pass through. It has been found, experimentally, that low-energy transfers are by far predominant; electrons must thus undergo large numbers of interactions, before coming to rest. The slowing down of electrons may therefore be treated as a gradual, continuous process, which may be characterized by the **linear energy transfer** (LET) to the material involved. LET is a quantity that is defined, at every point x of the path of a particle of energy $E(x)$, as the ratio of the energy lost by the particle, between points x and $x + dx$, divided by the path element dx. In radiation protection, we can approximate the **linear energy transfer** (LET), as the material's **linear stopping power, noted S**.

This quantity S, noted as dE/dx, is the measure of the amount of energy imparted by electrons to matter, per unit length of path traversed inside the material. LET values – usually expressed in keV·cm^{-1}, or in MeV·cm^{-1} – depend on the energy of the electrons involved, and the nature of the material they travel through.

Further information

Electron stopping power

The energy loss sustained by an electron, along its path, may effectively be broken down into two components: **collisional stopping power** (due to ionisation and excitation processes), and the stopping power component due to bremsstrahlung, viz. **radiative stopping power**.

Collisional stopping power

Collisional stopping power corresponds to the formula: $S_{col} = \left(\frac{dE}{dx}\right)_{col}$

A relation, known as the Bethe formula, provides an expression for the collisional stopping power, for electrons, as follows:

$$S_{col[keV \cdot \mu m^{-1}]} = 76 \cdot 10^{-4} \times \rho_{[g \cdot cm^{-3}]} \times \frac{Z}{A} \times \frac{1}{\beta^2}$$

$$\times \left[\ln\left(511 \cdot 10^6 \times \frac{\beta^2}{1-\beta^2} \times \frac{E_{e^-[keV]}}{\bar{I}^2_{[eV]}} \right) - \beta^2 \right]$$

2 – Interaction of ionising radiation with matter

Since the electron travels at relativistic speeds, the quantity $\beta = \frac{v_{e^-}}{c}$ is introduced, where v_{e^-} is the velocity of the electron, c the speed of light.

The parameters appearing in this expression are dependent either on the incident particle, or on the nature of the target material.

Parameters relating to the incident particle:
E_{e^-} stands for the electron's kinetic energy.

Parameters relating to the material:
ρ is the material's density;
Z is the material's atomic number;
A is the material's mass number;
I stands for the mean ionisation and excitation potential of atoms in the material, which increases with Z.

Radiative stopping power

Radiative stopping power may be noted as: $S_{rad} = \left(\frac{dE}{dx}\right)_{rad}$

The radiative stopping power can be readily calculated, in relation to the collisional stopping power:

$$\frac{S_{rad}}{S_{col}} = 1.4 \cdot 10^{-6} \times Z \times E_{e^-} [\text{keV}]$$

The **total stopping power** may then be written as:

$$S_{tot} = S_{col} + S_{rad}$$

With regard to the special case of living tissues, the "biological damage" caused by electrons is all the greater, the higher the amount of energy that is locally deposited in cells: LET thus stands as a crucial quantity, for the purposes of determining the "biological effectiveness" of radiation (see Chapter 4, "Biological effects of ionising radiation").

Figure 2.6 plots the variation of LET in tissue versus of electron energy.
It will be seen that:

- at low energies, LET decreases as energy rises. From this, it may be inferred that, the further electrons are slowed down, the higher the amount of energy they impart to matter through each interaction, and the greater their ability becomes to harm the constituent cells in tissue. Electrons thus present a greater hazard at the end of their path;

- in the range 500 keV < E < 5 MeV, LET may be regarded as effectively constant, with a value equal to about 2 MeV·cm^{-1}.

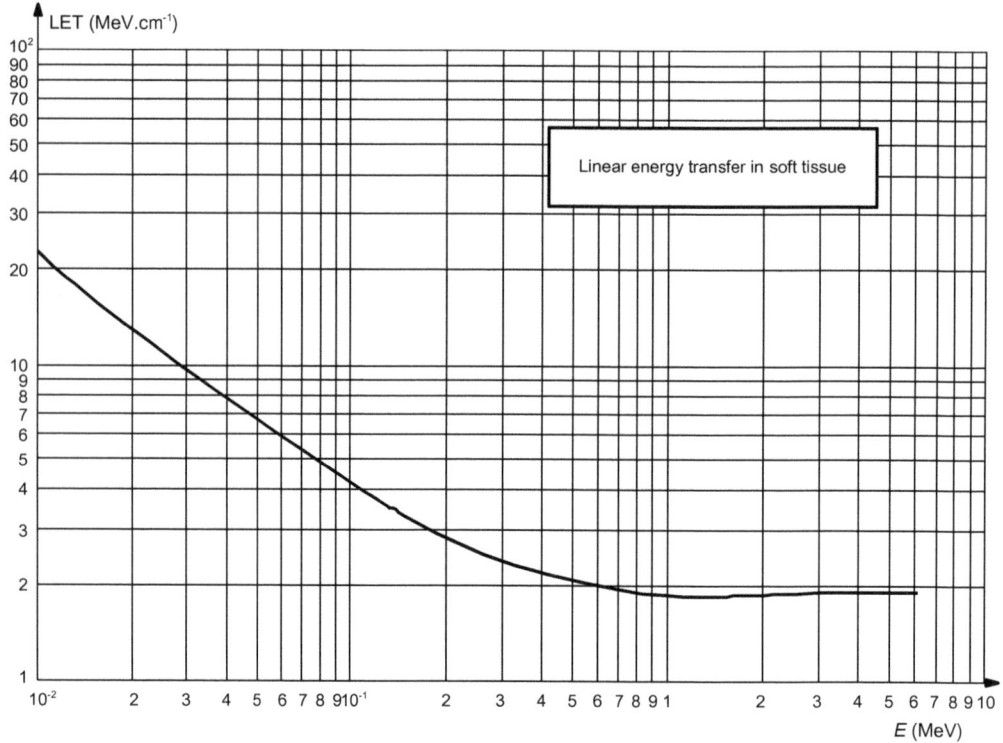

Figure 2.6. Variation of electron LET values in tissue as a function of electron energy.

2.2.2.5. Range of electrons in matter

Electrons follow highly tortuous paths: indeed, a deflection is found to arise, as a rule, for every interaction mode (ionisation, excitation, bremsstrahlung) that occurs. Some electrons may even undergo a 180° deflection (the backscatter phenomenon).

For the purposes of characterizing the paths of electrons, or beta radiation, two quantities may be defined (see Figure 2.7):

- **pathlength**: this is the actual distance traveled by the electron along its path. This quantity is not much used for radiation protection purposes;
- **range**: this corresponds to the maximum depth to which a given electron beam can penetrate a medium of interest. This quantity is widely used, in radiation protection work, for the purposes of designing shielding. "Pathlength" is sometimes – improperly – used, in general parlance, to refer to range.

Many empirical relations have been worked out, serving to calculate range values as a function of incident electron energy, according to the nature of the material traveled through. One such formulation, the Katz and Penfold formula, sets out the following relation:

$$R = \frac{0.412 E^n}{\rho} \quad \text{taking } n = 1.265 - 0.0954 \ln E$$

where R is the range (in cm), E electron energy (in MeV), ρ the material's density (in $g \cdot cm^{-3}$).

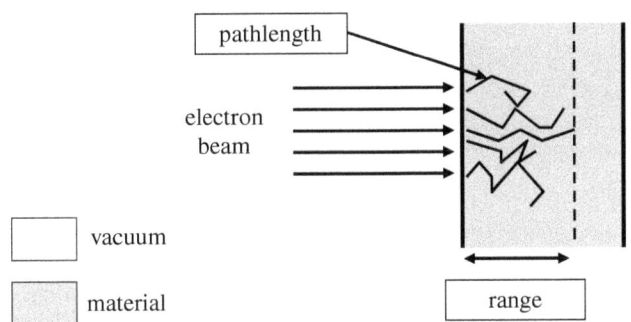

Figure 2.7. Pathlength and range of an electron beam.

For the special case of β radiation, with E substituted for by E_{max} – i.e. the characteristic maximum energy value for the spectrum of a given radiation – the formula set out above yields the maximum range, i.e. the range for the electrons of highest energy. While this maximum range may be at variance with actual physical occurrences (the average energy value for the β radiation spectrum is a closer fit for actual conditions), this quantity is widely used, for the purposes of shielding calculations. **For radiation protection purposes, it is usually deemed preferable to overestimate risk, so as to ensure adequate protection, rather than strive for a strict accuracy of results.**

As an illustration of the above relation, let us calculate the range of electrons endowed with an energy of 1.7 MeV (a value equal to the maximum energy of β particles emitted by phosphorus-32), in air and in water.

First, we can calculate the value of exponent n:

$$n = 1.265 - (0.0954 \times \ln 1.7) = 1.214$$

– in air, for which density ρ is taken to stand at $1.3 \cdot 10^{-3}$ g·cm^{-3}, the range is calculated to be equal to:

$$R_{air} = (0.412 \times 1.7^{1.214})/(1.3 \cdot 10^{-3}) = 604 \text{ cm} \qquad \text{(i.e. about 6 m);}$$

– in water, with density ρ equal to 1 g·cm^{-3}, the range is calculated to be equal to:

$$R_{water} = (0.412 \times 1.7^{1.214})/1 = 0.78 \text{ cm.}$$

β radiation thus has a **range of a few meters in air, and around 1 centimeter in soft tissue** (its density being close to that of water).

For most β$^-$-emitter radionuclides, the range for emitted β particles, in Plexiglas, stands at less than 1 cm.

If we now consider an average LET value, by dividing it by the initial (starting) electron energy, **a mean pathlength** can be calculated.

If $\bar{L}$ is the average LET value, $\bar{p}$ the mean path length, E_0 the initial energy of the electron, it is possible to write: $\bar{L} = \left(\overline{\frac{dE}{dx}}\right) = \frac{E_0}{\bar{p}}$, hence: $\bar{p} = \frac{E_0}{\bar{L}}$.

For electrons of 1.7 MeV starting energy, and an average $\bar{L}$ of 1.75 MeV·cm^{-1} (see Figure 2.6), the mean pathlength in soft tissue is calculated as follows.

Therefore: $\bar{p} = CSDA = \frac{E_0}{\bar{L}} = 1.7/1.75 = 0.97$ cm.

CSDA: "Continuous slowing down approximation range". The CSDA approximation is also associated with the notion of average value of the energy transferred by TEL.

Pathlength is thus seen to be longer than range. This shows the significant effect of deflections, with regard to electron paths.

2.2.2.6. The special case of positrons

All of the foregoing considerations apply equally to negatrons, and positrons alike. A significant, specific phenomenon does arise, however, concerning positrons, once they have lost all of their kinetic energy.

Once at rest, the positron comes together with a negatron, and the two particles vanish, their masses being converted into energy. This phenomenon is known as "**positron annihilation**", or "**matter–antimatter annihilation**". This is accompanied by the emission of two photons (quanta of electromagnetic energy) (see Figure 2.8).

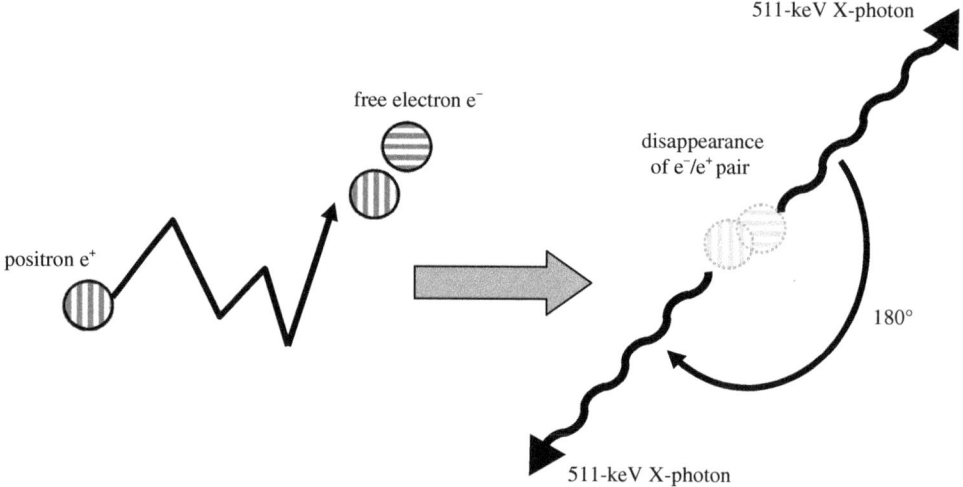

Figure 2.8. Positron annihilation process.

The general laws of conservation of energy, and momentum, when applied to annihilation, entail that these two X-photons, each with an energy of 511 keV, must be emitted in opposite directions (i.e. at an angle of 180°).

Any shielding serving to absorb positrons (β^+ radiation) thus turns into a secondary source of monoenergetic (511-keV) X-rays that are precisely twice as numerous as the absorbed positrons. From a radiation protection standpoint, it is imperative to take this phenomenon into account.

2.2.3. Interaction of heavy charged particles with matter: the case of alpha particles

For α particles, the slowing down processes involve **ionisations**, and **excitations** (see Figure 2.9). Alpha particles are regarded as heavy particles, compared to electrons, since they weigh in at some 8000 electron masses.

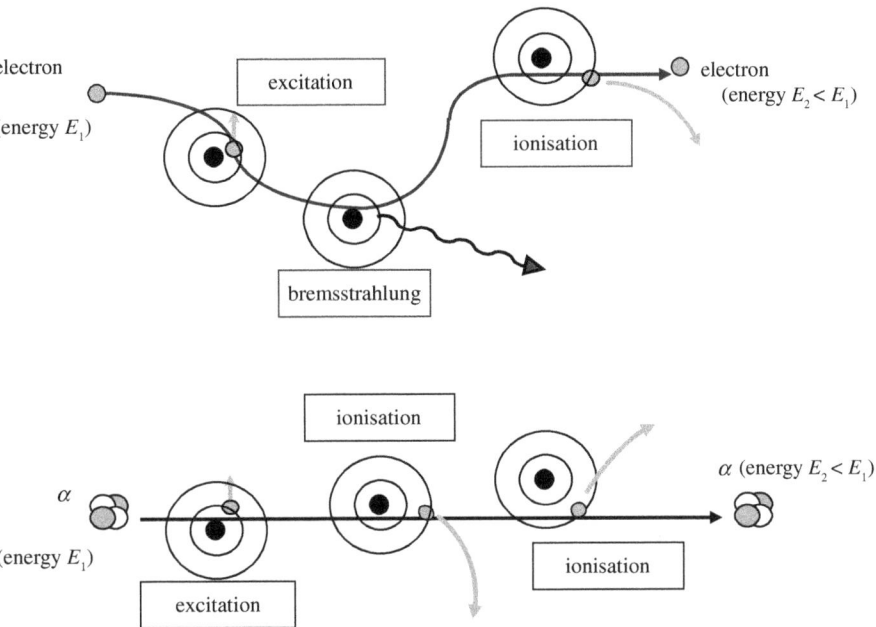

Figure 2.9. Comparison of electron and α-particle interactions in matter.

    The positive electric charge they bear, together with their higher mass than electrons, mean that the probability of interaction with atomic nuclei (i.e., bremsstrahlung), for α particles is quite low. They follow paths that are effectively straight, and quite short. Thus, for these particles, range and pathlength may be regarded as identical.

    The order of magnitude for the ranges of α particles, in air, stands at **a few centimeters** (see Table 2.1). In soft tissue, α-particle ranges are exceedingly short: of the order of a few tens of micrometers. Protection against this type of radiation is thus readily achieved, in the event of external exposure. Indeed, a mere sheet of paper acts as an adequate barrier to stop such particles, endowed with energies of around 6 MeV, as emitted by radioactive sources.

    As in the case of electrons, a definition may be given for LET, for such particles: viz., the average quantity of energy transferred, per unit length of path traversed.

    Taking an α particle endowed with an energy of 5 MeV, having a range in tissue of 50 μm, the LET value is found to stand at:

$$\text{LET} = 5/50 = 0.1 \text{ MeV} \cdot \mu\text{m}^{-1} = 100 \text{ keV} \cdot \mu\text{m}^{-1} = 1000 \text{ MeV} \cdot \text{cm}^{-1}$$

Table 2.1. Ranges in air, and water, for a few α particles.

Radionuclide	^{232}Th	^{210}Po	^{218}Po	^{212}Po
E_α (MeV)	4.2	5.3	6	8.8
R_α (cm) in air	2.6	3.8	4.6	8.6
R_α (μm) in water	32	48	57	107

Setting this result against the LET values found for electrons (around 2 MeV·cm^{-1}), it will be seen that **the energy imparted by alpha particles is 500 times larger** than for electrons, for a given pathlength. This type of radiation may thus be regarded as particularly harmful, should it directly impinge on living tissues.

Further information

It will be remembered that the linear energy transfer is equal to the loss of energy sustained by the incident particle, per unit length traversed. The only energy transfer mechanism involved, for heavy charged particles, is the collision process, by way of ionisation, and excitation. In contrast to what is the case for electrons, the stopping power (or LET) due to bremsstrahlung is nonexistent. The stopping power, for heavy charged particles – and thus for alpha particles – may therefore be calculated directly by means of the following form of the Bethe formula:

$$S_{[keV \cdot \mu m^{-1}]} = 14288 \times \frac{Z_{part}^2}{E_{k\,part\,[keV]}} \times \rho_{[g \cdot cm^{-3}]} \times \frac{Z_{mat}}{A_{mat}} \times \ln\left(\frac{2.195}{A_{part}} \times \frac{E_{k\,part\,[keV]}}{\bar{I}_{[eV]}}\right)$$

It will be noted that, since heavy particles are nonrelativistic, the term $\beta = \frac{v_{e^-}}{c}$, which appeared in the Bethe formula for electrons, is not included here.

The parameters appearing in this expression are dependent either on the incident particle, or on the nature of the target material.

Parameters relating to the incident particle:
$E_{k\,part}$ stands for the particle's kinetic energy;
Z_{part} is the particle's atomic number (for α particles: $Z_{part} = 2$):
A_{part} is the particle's mass number (for α particles: $A_{part} = 4$).

Parameters relating to the material:
ρ is the material's density;
Z_{mat} is the material's atomic number;
A_{mat} is the material's mass number;
$\bar{I}$ stands for the mean ionisation and excitation potential of atoms in the material, which increases with Z.

It follows from the Bethe formula that heavy charged particles sustain very high energy transfers, compared to electrons. The Bethe formula, when applied to alpha particles in water, yields the plot shown in Figure 2.10.

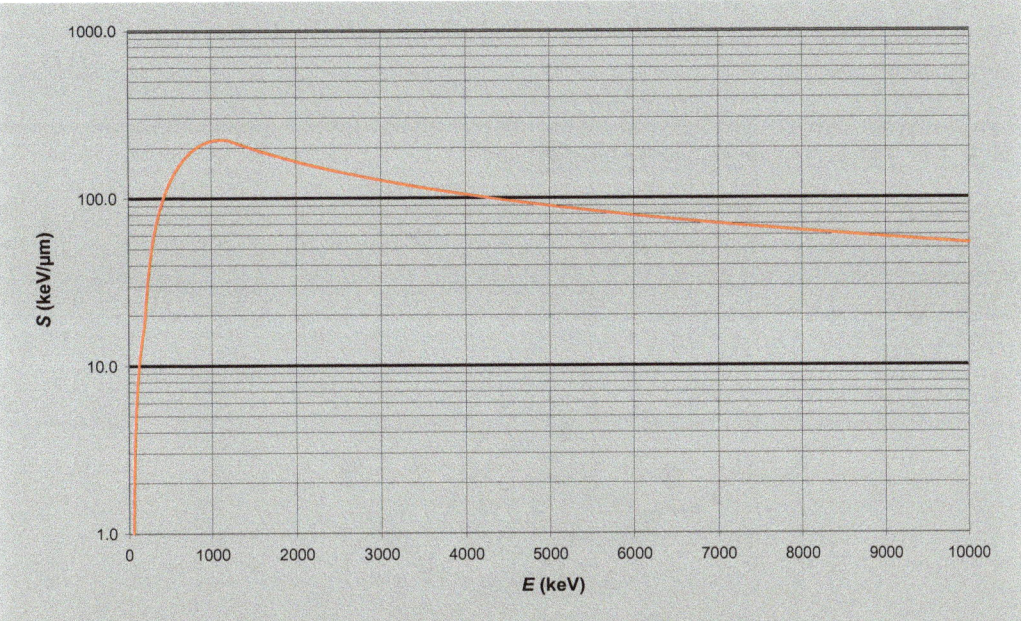

Figure 2.10. Linear stopping power for alpha particles in water (in keV/μm).

When a heavy charged particle enters a medium, its kinetic energy decreases in continuous fashion, while stopping power increases, fairly slowly at first, far more steeply towards the end of its path (this phenomenon shows up as the so-called **Bragg peak**).

2.3. Interaction of electromagnetic radiation with matter

Owing to its lack of any electric charge, electromagnetic radiation proves highly penetrating, in matter. This reflects the fact that the interactions of electromagnetic radiation with matter stand as "infrequent," "random" – i.e. "stochastic" – phenomena, compared to charged-particle interactions, which are commonly termed "deterministic". It follows that pathlengths, for such radiation, are extremely long: of the order of **several hundred meters in air**.

Electromagnetic radiation is also referred to as "indirectly ionising radiation". In effect, through its interactions, such radiation sets electrons in motion, which in turn ionize the material, by way of the phenomena reviewed in the preceding section.

For the range of energies we are concerned with here (from 0 to a few MeV), we may restrict our survey to the following interactions, arising between electromagnetic radiation and matter:

– the photoelectric effect;

– the Compton effect;

– pair production.

It should be noted that while X-rays, and γ-rays do have distinct, different origins, nevertheless, at comparable energies, their nature, and the interactions they give rise to, prove strictly identical.

2.3.1. The photoelectric effect

In this process, the incident radiation transfers all of its energy to an atomic electron in the material it is going through; the electron is thus ejected from its atom, with a certain quantity of kinetic energy (see Figure 2.11). Such an electron is also known as a **photoelectron**; its energy may be noted E_{pe}.

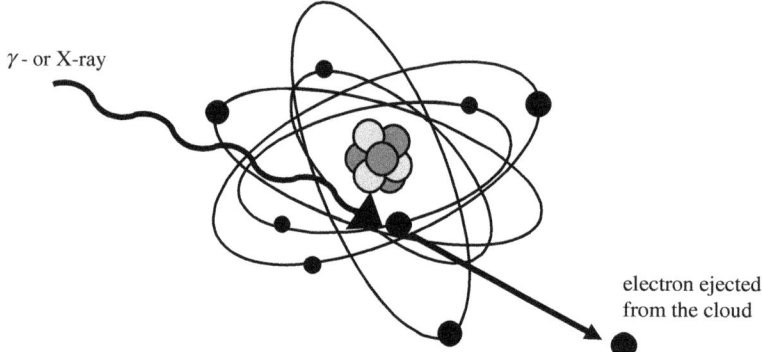

Figure 2.11. The photoelectric effect.

The value of this kinetic energy E_{pe} is given by the following relation:

$$E_{pe} = E_\gamma - E_b$$

where E_γ is the energy of the incident electromagnetic radiation (in this case, a γ-ray), E_b is the atomic electron's binding energy, in the electron shell it occupied.

For a given electron shell, the photoelectric effect can only occur if the energy of the incident photon is higher than the binding energy for that shell.

Provided the electromagnetic radiation involved does have sufficient energy, then, the higher the binding energy for an electron, the more likely it is this phenomenon will occur.

The photoelectric effect thus concerns tightly bound electron shells (K, or L shells). The atom thus undergoing ionisation in one of its inner shells ends up in an excited state. This results in a rearrangement of the electron cloud, with a concomitant emission of **X-rays**, and electrons (Auger electrons). The sum total of all the energies of the various radiations (particles included) issuing from this rearrangement is equal to the binding energy of the electron subject to the photoelectric effect.

The photoelectric effect may be seen as a single ionisation event. This liberates an electron endowed with sufficient energy to cause, all along its path across the material, thousands of ionisations, and excitations. It is the large numbers of such ionisations and excitations this event gives rise to that allow a signal to be detected, and measured. It follows that the actual energy of the gamma radiation is never directly measured, rather what is measured is the energy of the electron it has set in motion. Since the photoelectron carries off practically all of the gamma-ray energy, measuring that photoelectron's energy yields – indirectly – the gamma-ray energy. This causes spectrometry readouts to show a line spectrum, the lines corresponding to the energies of the photoelectrons produced.

The photoelectric effect is dominant at low energies. The probability of occurrence becomes much higher, for this effect, as incident photon energies tend to very low values. This probability of interaction is also strongly dependent on the atomic number Z of the material traversed. The heavier (higher-Z) the material is, the higher the ensuing probability of interaction. Which is why high-Z substances – e.g. lead – are the materials of choice, for the purposes of arresting γ or X-photons.

2.3.2. The Compton effect

The Compton effect likewise involves an interaction with an electron in the substance traversed by the incident radiation. It only occurs with electrons having a low binding energy, or free electrons.

In contrast to the photoelectric effect, in this case, the photon only imparts a fraction of its energy to the unbound, or loosely bound electron. The incident electromagnetic radiation's energy is distributed between the electron (referred to as a Compton electron, of energy noted E_{ce}) with which it has interacted, and a scattered electromagnetic radiation, known as the Compton-scattered photon (see Figure 2.12).

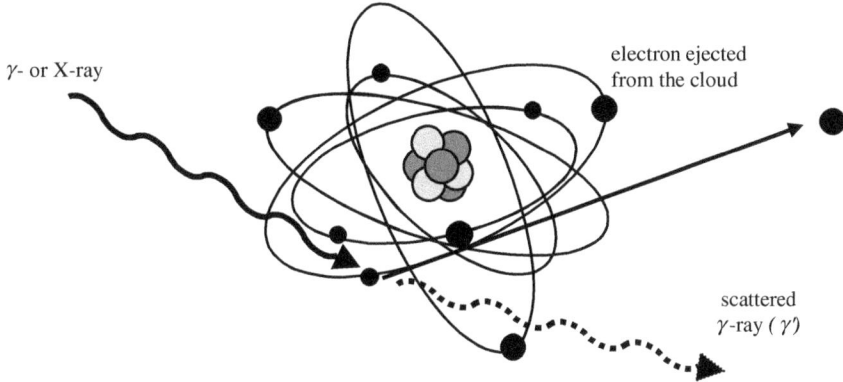

Figure 2.12. The Compton effect.

In the event of such an interaction due (say) to γ radiation, the value for the Compton electron's kinetic energy E_{ce} is given by the following relation:

$$E_{ce} = E_\gamma - E_{\gamma'}$$

where E_{ce} is the kinetic energy of the Compton-ejected electron, E_γ the energy of the incident electromagnetic radiation, $E_{\gamma'}$ the energy of the scattered electromagnetic radiation.

This phenomenon – which may be regarded as a scattering process – is all the more likely to occur, the lower the binding energy is, for the atomic electron involved; in contrast to the photoelectric effect, the Compton effect concerns electrons in loosely bound shells. This is why the electron's binding energy E_b is omitted, in the above formula.

Further information

Energy balance of the Compton effect

As stated above, in this interaction, the photon only imparts a fraction of its energy to the unbound, or loosely bound electron. The energy not thus transferred remains in the form of a photon, with a different energy, and direction of travel, compared to the incident photon's energy and direction. Depending on the quantity of energy transferred, many possibilities exist, as regards the scattered photon's direction, at an angle (noted θ) which may vary from 0° to 180° (see Figure 2.13).

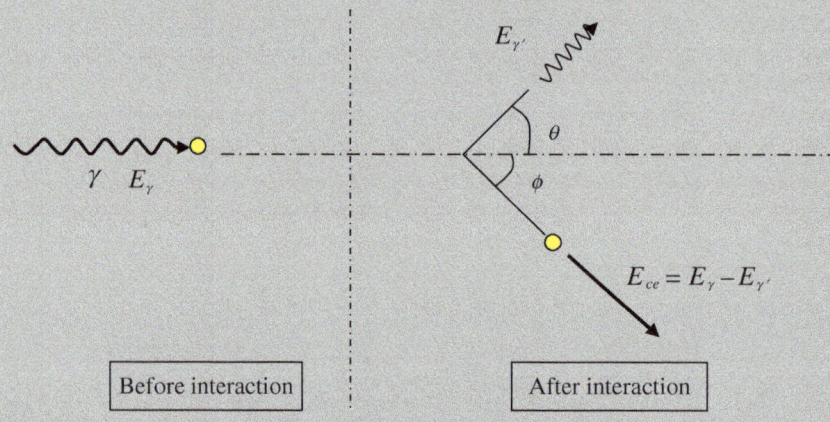

Figure 2.13. Geometry of the Compton effect.

For the purposes of determining the energy of the scattered photon, and of the Compton electron, the process is regarded as an elastic interaction, complying in principle with the conservation of momentum during the collision.

The conservation of energy may then be written as follows:

$$E_{\gamma'} = \frac{E_\gamma}{1 + \frac{E_\gamma(1-\cos\theta)}{m_0 c^2}}$$

where $E_{\gamma'}$ is the energy of the scattered photon;
E_γ is the energy of the incident photon;
$m_0 c^2$ is the rest mass-energy of the electron;
θ is the scattering angle, between the incident and the scattered photons.

It is found that, for every value of angle θ, there corresponds a value for the energy of the scattered photon, $E_{\gamma'}$. Thus, the Compton-scattered photon's energy may vary from $E_{\gamma'} = E_\gamma$, for an angle $\theta = 0°$ (no interaction), to a minimum value, for $\theta = 180°$. The latter case corresponds to a backscatter event.

Taking a rounded value for the electron rest mass-energy, viz. 0.5 MeV, the minimum scattered photon energy, i.e. for the backscattered photon, may be calculated by means of the following approximate formula:

$$E_{\gamma min}[\text{MeV}] = \frac{E_\gamma[\text{MeV}]}{1 + 4E_\gamma[\text{MeV}]}$$

The electron's kinetic energy, $E_{ce} = E_\gamma - E_{\gamma'}$, is thus zero for an angle $\theta = 0°$ ($E_{\gamma'} = E_\gamma$), and reaches a maximum $E_{ce\ max}$ for an angle $\theta = 180°$ (when $E_{\gamma'}$ is at its minimum value). In interactions of this type, the electron thus never carries off the full energy of the incident gamma-ray. As a result, with the Compton effect, the energy spectrum, for Compton electrons, is a continuous spectrum, ranging from 0 to $E_{ce\ max}$, depending on the scattering angle. On the other hand, the scattered photon may also interact within the detector, releasing its energy by way, e.g., of a photoelectric effect. In that case, the total energy released in the detector is equal to the energy of the Compton electron, $E_{ce} = E_\gamma - E_{\gamma'}$, plus the energy of the photoelectron, effectively $E_{pe} = E_{\gamma'}$. The Compton electron and the photoelectron, in conjunction, will thus deposit, by way of ionisations and excitations, a quantity of energy equal to the full energy of the incident photon. This full energy will show up in the spectrum, as a gamma-ray line.

For the Compton interaction, the probability of interaction is dominant for photons of intermediate energies, while remaining far from insignificant in the low- and high-energy ranges. It is also dependent on the atomic number Z of the material traversed.

2.3.3. Pair production

Pair production involves interactions with atomic nuclei in the material the electromagnetic radiation passes through. A gamma photon interacts with the electric field of a nucleus, and disappears, yielding a pair of particles, consisting of one electron, and one positron. This pair production phenomenon is also referred to as a "materialization" effect (see Figure 2.14).

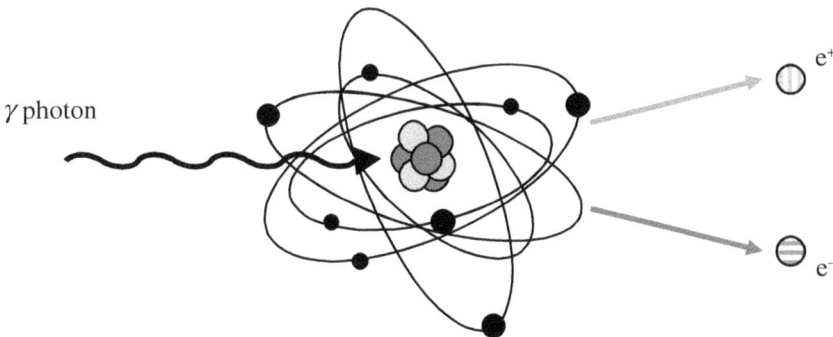

Figure 2.14. Pair production.

To comply with the law of energy conservation, the incident electromagnetic radiation must have an energy at least equal to the energy equivalent of two electron masses, if such materialization is to be possible: i.e. 1.022 MeV. Below this value, occurrence of this effect is impossible, in energy terms. At energies above this threshold value, any excess energy takes the form of kinetic energy – which may be noted E_{pp} – imparted to the electron–positron pair:

$$E_{pp} = E_{e^-} + E_{e^+} = E_{\gamma\,[MeV]} - 1.022 \text{ MeV}$$

where E_{e^-} is the kinetic energy of the electron, E_{e^+} the kinetic energy of the positron.

This kinetic energy is not necessarily evenly distributed between the electron, and the positron. Any combination can, and does occur. For instance, the electron may receive 30% of the available kinetic energy, the positron 70%, or vice versa.

It will be remembered that, as was pointed out in section 2.2.2.6, the positron generated by this interaction ultimately disappears, at the end of its path, when it comes together with an atomic electron. The resulting annihilation yields two photons, each with an energy of 511 keV.

In spectrometry readouts, this process shows up as a gamma-ray line corresponding to the total kinetic energy of the electron–positron pair, this being equal to $E_{\gamma\,[MeV]}$ – 1.022 MeV. On the other hand, one annihilation photon (0.511 MeV), or both photons (1.022 MeV) may also be absorbed, giving rise to two further lines.

In contrast to what is the case for the photoelectric effect, the probability of occurrence for pair production, above the threshold value for that interaction, i.e. 1.022 MeV, increases, as incident photon energies tend to very high values. This probability of interaction is also dependent on the atomic number Z of the material traversed. The heavier (higher-Z) the material, the higher the probability of pair production becomes.

It will thus be seen that electromagnetic radiation, by way of the three effects outlined above, generate (negatron–positron pair), or set in motion (photoelectron, Compton electron) electrons, which in turn ionize the material. This is why such radiation is termed indirectly ionising radiation.

2.3.4. Attenuation of electromagnetic radiation

2.3.4.1. Attenuation coefficient

In contrast to charged particles, which gradually transfer their energy to matter by impinging on thousands of electrons along their path, electromagnetic radiation comes to an end, as a rule, in one sudden event, by way of a single interaction. Further, the concepts of slowing down, or pathlength, appropriate as they are for charged particles, are not applicable, for such radiation. Another quantity is used, for the purposes of evaluating the interactions of electromagnetic radiation in matter: **the concept of attenuation** in terms **of photon numbers**.

Consider a uniform, parallel beam of monoenergetic electromagnetic radiation, penetrating into a shield of thickness x (see Figure 2.15). The fluence at beam entry, i.e. the number of photons impinging on the shield, per unit surface, is taken to be ϕ_0 (fluence is expressed in cm^{-2}, since a number – e.g. of photons – has no specific unit).

It can be shown that the fluence ϕ (i.e. the number of photons emitted, per unit surface) for the monoenergetic radiation exiting the shield, along the initial beam direction, and

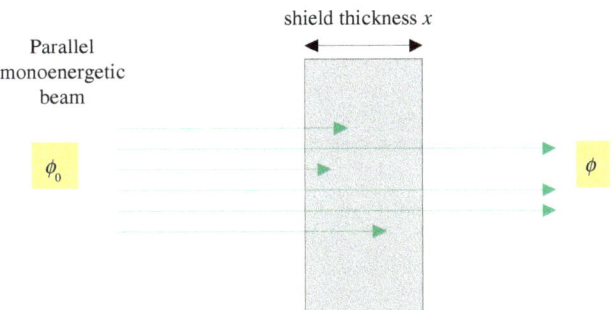

Figure 2.15. Attenuation of radiation across a shield.

retaining the initial energy, i.e. the number of photons that have undergone no interaction while passing through the shielding material, is given by the following relation:

$$\phi = \phi_0 e^{-\mu x}$$

where μ is the linear attenuation coefficient, which is expressed as the reciprocal of distance: if x is expressed in cm, μ must be expressed in cm^{-1}. This **attenuation coefficient** is a measure, in effect, of the probability of interaction with matter, for this radiation. A high μ coefficient indicates a strong attenuation of radiation in the material of interest. In other words, the higher the value for μ, the stronger the photon interactions are found to be. This coefficient is dependent on the energy of the radiation involved, and on the nature of the shielding material.

This attenuation relation may equally serve to calculate the number N of photons that emerge without having undergone any interaction, after passing through a shield of thickness x, by reference to the number N_0 of incident photons:

$$N = N_0 e^{-\mu x}$$

when plotted on semilogarithmic graph paper, this attenuation law yields a straight line, of slope $-\mu$.

It should be noted that this law can only serve to calculate the number of photons that emerge from the shield while retaining the initial energy of the incident beam (i.e., photons that have not been involved in any interaction). Now, such photons only account for a fraction of emerging photons: they do not include, in particular, the Compton-scattered photons.

Let's note that for physical and educational reasons, quantity $1/\mu$, called mean free path of photons, is often preferred to the linear attenuation coefficient μ because it allows more easily to measure the problems of radiation protection.

The attenuation coefficient μ stands as the sum of partial attenuation coefficients, relating to each of the effects identified (photoelectric effect, Compton effect, pair production):

$$\mu = \mu_{\text{pe}} + \mu_C + \mu_{\text{pp}}$$

where μ_{pe} is the photoelectric attenuation coefficient, μ_C the Compton attenuation coefficient, μ_{pp} the pair production attenuation coefficient.

It should be noted that some authors prefer to use the following notation: $\mu = \tau + \sigma + \kappa$; τ, σ, κ being the photoelectric attenuation coefficient, the Compton attenuation coefficient and the pair production attenuation coefficient, respectively.

The value of each of these coefficients is dependent, just as the total attenuation coefficient μ, on:

- the nature of the target material;
- the energy of the incident radiation.

Finally, the relation $\phi = \phi_0 e^{-\mu x}$ allows three fundamental parameters to be derived, for the purposes of photon shielding calculations:

- **the transmission factor**, i.e. the ratio ϕ/ϕ_0, expressing the percentage of photons that get through the shield;
- conversely, the ratio ϕ_0/ϕ, with a value higher than unity, viz. the **attenuation factor**;
- finally, the ratio $(\phi_0 - \phi)/\phi$, or $1 - \phi/\phi_0$, a percentage measuring photon **attenuation** within the shield.

We shall encounter repeatedly these important parameters, as we perform shielding calculations for the purposes of protecting against external exposure to X- and γ-radiation.

2.3.4.2. Domains of prevalence for electromagnetic radiation interaction mechanisms

It should be noted that the probability of occurrence, for any one of these three effects (photoelectric effect, Compton effect, pair production), depends on the **energy of the electromagnetic radiation** involved, and on the **nature of the material** through which this radiation travels.

The plots appearing in Figures 2.16 and 2.17 show how attenuation coefficients vary with energy in soft tissue (the result is practically the same as in water), and lead. Photon energy (in MeV) is plotted along the x-axis, the attenuation coefficient (in cm^{-1}) along the y-axis.

The prevalence of each of these effects is seen to be consistent with what was said in the preceding section.

The photoelectric effect stands as the dominant effect at low energies. At the same time, as may be noted in the case of lead, discontinuities are seen to arise for incident γ-ray energies corresponding to the binding energies for the K and L shells. To understand this phenomenon, we may consider, for instance, the K shell in lead, for which the binding energy strands equal to 88 keV. It will be remembered that the photoelectric effect can only occur, for a given electron shell, if the gamma photon is of higher energy than the binding energy for that shell. In this case, if the gamma photon's energy is less than 88 keV, photoelectric effects do remain possible, involving any shell other than the K shell. On the other hand, should the gamma photon's energy be slightly higher than 88 keV, then photoelectric effects become possible for all electron shells (including the K shell), resulting in an abrupt increase in the probability of interaction for the effect. Such a discontinuity arises for every binding energy value, for the K, L, M... shells of the various constituent

2 – Interaction of ionising radiation with matter   63

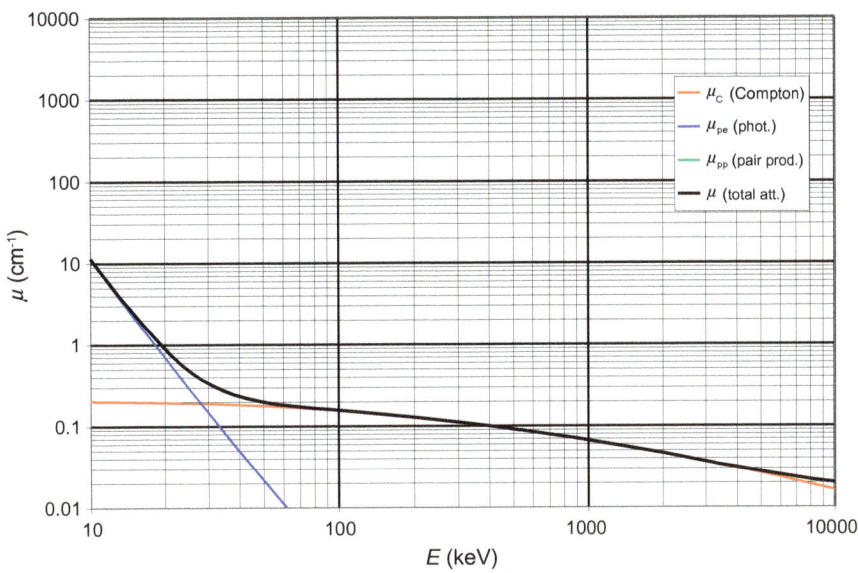

Figure 2.16. Attenuation coefficient in tissue.

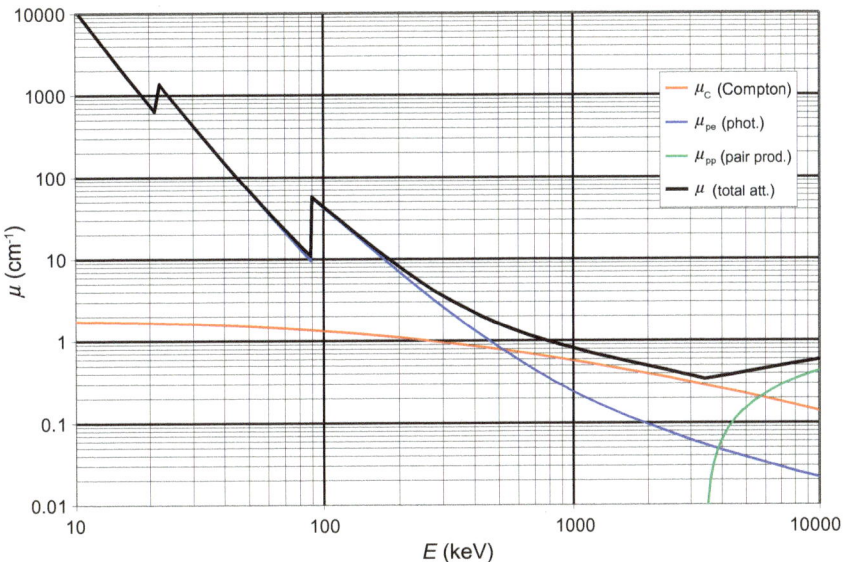

Figure 2.17. Attenuation coefficient in lead.

materials, of all kinds, in a shield. No such discontinuity appears in Figure 2.16, since the binding energies for the K, L, M… shells all stand lower than 10 keV for tissue, and concrete, whereas the γ energy scale used starts at 10 keV.

The Compton effect is dominant at intermediate energies, once the photoelectric effect becomes negligible, and up to the point where pair production becomes significant. For low-Z materials, e.g. tissue, or water, this effect is practically the only one that need be taken into account. Its presence is apparent across an extensive energy band, ranging from 80 keV to 10 MeV.

Pair production is the dominant effect at high energies. It is significant only in high-Z materials, and for energies higher than several MeV.

When calculating attenuation, all three types of effects liable to occur must be taken into account. The total attenuation coefficient is used, this being equal to the sum of the three individual coefficients: it appears as a heavy line in the plots shown here. It will be seen that this attenuation coefficient is all the higher, the higher the atomic number of the material (i.e. for high-Z materials). Thus, for a 1-MeV gamma-ray, the attenuation coefficient is found to be 0.07 cm^{-1} in soft tissue, 0.80 cm^{-1} in lead.

Exercises

Consider a 1-MeV electromagnetic radiation beam impinging on a lead shield, with a fluence ϕ_0 of 1 000 000 photons/cm^2. From Figure 2.17, μ is found to be equal to 0.80 cm^{-1}.

1. *Calculate the fluence ϕ of the radiation transmitted through a 1-mm lead shield.*

$\phi = \phi_0 e^{-\mu x}$ therefore: $\phi = 1\,000\,000 \times e^{-0.80 \cdot 0.1} = 923\,116$ photons/cm^2.

Photons are still emerging from the shield in very large numbers. 923 116 photons/cm^2 are found, rather than 1 000 000 photons/cm^2, i.e. a transmission factor of 92.3%. Since the attenuation coefficient cannot be raised or modified (or only by switching to a shield of a material of higher Z than lead), in order to achieve a sharper reduction in the number of photons emerging from the shield, we can increase the thickness of our lead shield.

2. *Calculate the fluence ϕ of the radiation transmitted through a 5-cm lead shield.*

$\phi = \phi_0 e^{-\mu x}$ therefore: $\phi = 1\,000\,000 \times e^{-0.80 \cdot 5} = 18\,315$ photons/cm^2.

Going for a 5 cm thick lead shield results in only 1.8% of photons getting through. The attenuation factor ϕ_0/ϕ may likewise be calculated: this is found to be equal to 56. Fluence is thus attenuated by a factor of 56. Attenuation within the shield $(\phi_0 - \phi)/\phi$ may also be calculated: this is found to stand at 98.2%. In other words, 98.2% of the incident fluence is absorbed in the shield.

It can be seen that, by increasing shield thickness, a sharp reduction is achieved in the number of photons emerging – albeit without totally arresting them. There is always a probability – however small it may turn out to be – that some photons will emerge from the shield. Any calculation must take this possibility on board. A tradeoff is sought, such that a shield sufficiently thick to block very large numbers of gamma-rays is arrived at, without getting exceedingly large, so that bulk, mass, cost... can be kept within reasonable bounds.

2.3.4.3. Half-value and tenth-value layers

In radiation protection work, for the purposes of speeding up calculations, **half-value layers** (HVL), or **tenth-value layers** (TVL) are often considered, i.e. shield thickness values such that the fluence of the emerging radiation is equal, respectively, to one half, or one tenth of the fluence (**half attenuation, attenuation to one tenth**) that would obtain without such a shield.

By definition, half attenuation may be written as:

$$\frac{\phi}{\phi_0} = e^{-\mu x_{1/2}} = 1/2$$

therefore: $\dfrac{\phi_0}{2} = \phi_0 e^{-\mu x_{1/2}}$

and: $e^{\mu x_{1/2}} = 2$

Consequently, the half-value layer is equal to:

$$x_{1/2} = \frac{\ln 2}{\mu}$$

Likewise:

$$x_{1/10} = \frac{\ln 10}{\mu}$$

Examples:

1. *Calculate the half-value layer, and the tenth-value layer for 1-MeV electromagnetic radiation traversing a lead shield.*

It will be remembered that, in this case, μ is equal to 0.80 cm^{-1}.
Therefore:

$$x_{1/2} = \frac{\ln 2}{0.80} = 0.87 \text{ cm}$$

$$x_{1/10} = \frac{\ln 10}{0.80} = 2.88 \text{ cm}$$

2. *Calculate the number of half-value, and tenth-value layers that must be positioned in front of 1-MeV electromagnetic radiation so that just 5% is let through.*

If 1 tenth-value layer is interposed, the result is one tenth of incident radiation fluence, i.e. 10%.

If 1 half-value layer is further inserted, outgoing radiation will be reduced to one half of 10%, i.e. 5% of incident radiation.

In this case, the lead shield to be interposed consists of one tenth-value layer, and one half-value layer. The total lead thickness required is therefore equal to:

$$x_{total} = x_{1/2} + x_{1/10} = 0.87 + 2.88 = 3.75 \text{ cm}$$

2.3.4.4. Buildup factor

As was stated above, the fluence, for electromagnetic radiation emerging from a shield, is calculated by means of the following relation:

$$\phi = \phi_0 e^{-\mu x}$$

Now, in many instances, the actual fluence value, for the emerging electromagnetic radiation, is found to be higher than the value calculated by means of the above relation: indeed, the three effects involved (photoelectric effect, Compton effect, pair production) have the ability to yield – directly or indirectly – further photons, in particular Compton-scattered photons. Should such secondary photons emerge from the shield, this results in an increased fluence. These secondary photons involve lower energies, and propagate along different directions, compared to the initial beam.

The actual fluence value found, at any given point behind a shield, thus involves two components: one, which may be calculated as above, concerns those photons that have not undergone any interaction, and retain the initial energy, and direction (the above relation is used); the second component, which is far less readily evaluated, relates to all those emerging photons that retain neither the energy, nor the direction of the initial beam. In effect, the value for the latter component depends on:

- the nature, shape, and thickness of the shield;
- the energy of the incident radiation;
- the position of the point for which fluence is to be determined.

In order to take such secondary emerging photons into account, a buildup factor B is introduced into the relation set out above. This now reads:

$$\phi = B\phi_0 e^{-\mu x}$$

This buildup factor B, which is greater than unity, has the effect of increasing the fluence value, for emerging photons. Not taking this factor on board, in shielding calculations, would lead to underestimating fluence. For radiation protection purposes, as the buildup factor is not readily amenable to calculation, to speed up calculations, slightly larger values are adopted, for half-value and tenth-value layers, compared to the calculated values. To carry out more exacting calculations, for complex installations, buildup factor calculation codes are used.

2.4. Interaction of neutrons with matter

2.4.1. General considerations

The neutron is a particle bearing no electric charge, with a mass close to that of the proton. It is unstable when in the unbound state, with a half-life of 10 minutes.

Neutrons are generally classified according to their energy. This classification may be summarized in terms of the following table (Table 2.2).

Neurons interact only with atomic nuclei, in the materials they pass through. These interactions fall into two categories: those that result in the neutron disappearing, which are referred to as **absorption** events (capture, or fission events), and those that only result in a reduction in the neutron's energy, referred to as **scattering**.

Table 2.2. Classification of neutrons according to kinetic energy.

Neutron type	Kinetic energy
thermal	less than 0.4 eV
intermediate	from 0.4 eV to 200 keV
fast	from 200 keV to 10 MeV
relativistic	higher than 10 MeV

2.4.2. Neutron absorption

A neutron absorption event involves the neutron penetrating the target nucleus. Two situations may arise. Either the resulting compound nucleus, after a brief lifespan, yields some form of radiation – a γ-ray, an α particle, a proton, etc.: **this constitutes a capture event**. Or the nucleus splits into two or more fragments: **this is a fission event**. Regardless of the type of radiation emitted, or of the fragments produced, this process is also referred to as a **nuclear reaction**. The most common nuclear reaction encountered is radiative capture, noted (n, γ). Subsequent to emission of radiation, or after fissioning, the resulting nucleus, or nuclei are commonly radioactive, being found, as a rule, to be β emitters.

Examples of absorption reactions in matter:

$$^{4}_{2}He + ^{1}_{0}n \rightarrow ^{3}_{1}H + ^{1}_{1}p$$

$$^{6}_{3}Li + ^{1}_{0}n \rightarrow ^{3}_{1}H + ^{4}_{2}\alpha$$

$$^{59}_{27}Co + ^{1}_{0}n \rightarrow ^{60}_{27}Co + \gamma$$

$$^{235}_{92}U + ^{1}_{0}n \rightarrow 2FPs + (2-3)n$$

The last example corresponds to the neutron-induced fission reaction in uranium-235 nuclei; "FPs" stands for "fission products". It should be noted that this phenomenon is accompanied by the release of very large quantities of heat (see Box "Further information" below).

The probability, for a neutron, that it will undergo absorption is inversely proportional to its velocity. It will thus stand at a maximum for the slower neutrons, known as thermal neutrons, as their velocity is comparable to the thermal agitation in the medium they are traversing (i.e. with an energy of around 0.025 eV at a temperature of 25 °C). At the same time, the probability of neutron absorption, for some substances (cadmium, boron), may become very high for certain particular neutron energy values. As a rule, for fast neutrons to be absorbed, they must first be slowed down, for instance by passing through a hydrogenous material. Thus, the usual practice, to ensure protection against neutrons, is to make use of shielding involving a number of materials, mixed or in layered ("sandwich") form. The first material (the "moderator") serves to slow down neutrons, while the second one is a neutron absorber, which becomes effective, once the neutrons have been slowed down. A third shield may further be added, made of lead, to block the greater part of the γ radiation yielded by radiative nuclear reactions.

In practice, it may prove convenient to make use of one and the same material, e.g. water, for the purposes both of slowing down ("moderating"), and absorbing neutrons. This is commonly found to be less effective than a shield consisting of two fully dedicated materials (a moderator, and an absorber).

Further information

The nuclear fission reaction (from *CEA Thematic Booklet No. 5 – Nuclear Energy: Fission and Fusion*)

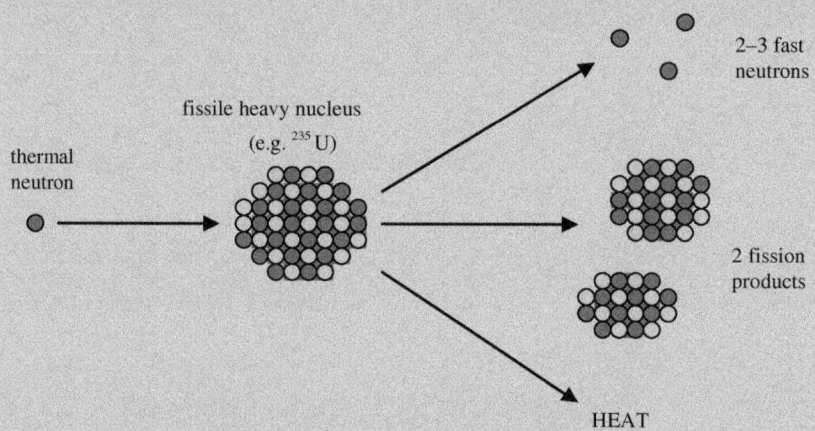

Figure 2.18. Principle of the fission reaction.

The chain reaction

Fission involves the breakup of a large nucleus (e.g. a uranium-235 nucleus), which, as result of a neutron impinging on it, splits into two smaller nuclei. This fission event is accompanied by the release of a large quantity of energy. It further involves the emission, simultaneously, of two, or three neutrons.

The neutrons thus released can in turn cause further fissions, in other nuclei, thereby releasing further neutrons, and so on. A chain reaction is thus set up, since, by initiating a single fission in the uranium mass, if the ensuing neutrons are not controlled, at least 2 further fissions are obtained, which in turn cause 4 fissions, then 8, 16, 32…

The two chief applications of nuclear fission are nuclear reactors, and nuclear weapons ("atom-bombs"). In nuclear reactors, the chain reaction is stabilized to keep it at a particular level: in other words, a major part of the neutrons is made to undergo capture, to prevent them from causing further fissions. All that is required is that one neutron, after each fission, is left to cause a further fission, thus releasing energy at each step. […] In contrast, in a nuclear weapon, the chain reaction must diverge as extensively as possible, within as short a time as feasible: its exponential rise is promoted, while energy is kept confined as long as possible.

> **Critical mass**
>
> Neutrons may be captured by certain atomic nuclei (e.g. uranium-238 nuclei, present in the uranium mass together with uranium-235) or indeed escape – thus causing no further fissions. For the chain reaction to be set up, a sufficient mass of fissile nuclei must thus be brought together in a given volume: this is known as the "critical mass".
>
> Critical mass is a crucial parameter in military applications of nuclear energy, for which, in contrast to civilian applications, the chain reaction must propagate extremely fast, in unrestricted fashion.

2.4.3. Neutron scattering

Scattering involves a collision, taking place between a neutron, and a nucleus.

Scattering is said to be elastic when it involves a neutron–nucleus collision that may be likened to the impact of a moving rigid ball on another, immobile rigid ball. The study of this type of impact shows that the loss of energy, for the ball acting as a projectile, reaches a maximum when its mass equals that of the target ball. In order to ensure effective slowing down of neutrons, the materials used should thus consist of light (low-Z) atoms, involving nuclei with masses approaching that of the neutron: the best moderator is thus hydrogen. This is why hydrogenous (hydrogen-rich) materials are used, to ensure effective slowing down of neutrons, requiring a minimum number of impacts, i.e. of scattering events. Mention may be made, for instance, of such materials as water, paraffin, concrete. Elastic scattering is the dominant process for neutrons of energies ranging from 1 keV to 500 keV.

Scattering is said to be inelastic if the target nucleus, subsequent to impact, is brought to an excited state, resulting in the emission – as a rule immediately (prompt emission) – of one or more γ-rays. Inelastic scattering is the dominant phenomenon for neutrons with energies higher than 500 keV. Just as electromagnetic radiation, neutrons are a form of indirectly ionising radiation: indeed, it is the radiations (including particles) yielded by the nuclear reactions resulting in neutron absorption that go on to be the main cause of ionisation in the material.

2.4.4. Neutron attenuation

As is the case for electromagnetic radiation, a parallel, uniform beam of monoenergetic neutrons undergoes exponential attenuation, depending on the shield thickness it passes through. As for γ- and X-rays, this exponential relation may only serve to calculate the neutron fluence, or the number of neutrons that emerge from the shield while retaining the energy, and direction of the initial beam – even as other neutrons, involving lower energies, and other directions also emerge. A similar expression as for γ-ray fluence is found for neutron fluence, given by the following relation:

$$\phi = \phi_0 e^{-\Sigma x}$$

where ϕ_0 is the initial fluence, ϕ the fluence after attenuation, x the shield thickness, Σ the **macroscopic cross-section**, which is expressed as the reciprocal of distance: if x is expressed in cm, Σ must be expressed in cm^{-1}. This coefficient closely resembles the attenuation coefficient μ for electromagnetic radiation. In neutronics, however, the preferred usage is to refer to this quantity as Σ, rather than μ.

Moreover, just as for photons with $1/\mu$, for neutrons, the mean free path $1/\Sigma$ is often preferred Σ for practical reasons.

Another quantity is also used to express the probability of interaction, for neutrons in a medium. This is the **cross-section**, noted σ. As indicated in the preceding chapter, this may be seen as a *notional surface area*, presented by each atom, such that, should a neutron hit the atom (i.e. the atomic nucleus) within this target area, the effect of interest will occur. Cross-sections are expressed in terms of units of surface area (cm^2). As cross-sections are very small – of the order of the actual cross-sectional area of a nucleus – a more practical unit is adopted, to measure cross-sections: the barn (symbol: b). 1 barn is equal to 10^{-24} cm^2. The macroscopic cross-section Σ is linked to the (atomic) cross-section σ by way of the following relation:

$$\Sigma(\text{cm}^{-1}) = n_c(\text{cm}^{-3}) \cdot \sigma(\text{cm}^2)$$

where n_c is the number of atoms per cubic centimeter within the shield. Should all the target atoms be identical, then, if A is the material's molar mass; the following relation obtains:

$$n_c = \frac{\rho}{A} N_A$$

where ρ is the target material's density, expressed in g·cm^{-3}, and N_A is Avogadro's number, equal to $6.022 \cdot 10^{23}$ atoms per mole.

In many texts, the probability of interaction is expressed in terms of the cross-section σ. Consequently, the relation giving the fluence value then reads:

$$\phi = \phi_0 e^{-\sigma n_c x}$$

To round off this review of cross-section concepts, it should be noted that the probability of interaction, for a neutron, with the material it is passing through is expressed in terms of a total cross-section σ_{tot}, which is equal to the sum of the probabilities for each of the various types of event liable to occur, as expressed by the respective partial cross-sections:

$$\sigma_{tot} = \sigma_{absorption} + \sigma_{scattering} \quad \text{(usually noted as: } \sigma_t = \sigma_a + \sigma_s\text{)}$$
$$\text{with: } \sigma_{absorption} = \sigma_{capture} + \sigma_{fission} \quad \text{(usually noted as: } \sigma_a = \sigma_c + \sigma_f\text{)}$$

The absorption cross-section is dominant for thermal neutrons, whereas the scattering cross-section dominates for fast, or intermediate-energy neutrons.

Examples:

Consider a beam of 1-MeV neutron radiation, resulting in a fluence of 1 000 000 neutrons per cm^2. To ensure some protection, the source is placed in a 10-centimeter-deep water basin.

1. *Calculate the number n_c of hydrogen atoms, and oxygen atoms per cubic centimeter of water.*

First, we can calculate the number of water molecules per cubic centimeter. Water density stands at 1 g·cm^{-3}, while the molar mass of water is 18 g.

$$n_{c\ water} = \frac{\rho}{A} \times (6.022 \cdot 10^{23}) = 3.345 \cdot 10^{22} \text{ molecules of water per cubic centimeter}$$

We recall that the water molecule consists of two atoms of hydrogen, for one atom of oxygen. Therefore:

$$n_{c\ hydrogen} = 3.345 \cdot 10^{22} \times 2 = 6.690 \cdot 10^{22} \text{ atoms of hydrogen per cm}^3$$
$$n_{c\ oxygen} = 3.345 \cdot 10^{22} \times 1 = 3.345 \cdot 10^{22} \text{ atoms of oxygen per cm}^3$$

2. Bearing in mind that the capture cross-sections, for 1-MeV neutrons, are equal to 4.52 barns, and 3 barns, respectively, in hydrogen, and in oxygen, calculate the macroscopic cross-sections in hydrogen (Σ_H), in oxygen (Σ_O, and the total macroscopic cross-section (Σ_{tot}).

$$\Sigma(\text{cm}^{-1}) = n_c(\text{cm}^{-3}) \cdot \sigma(\text{cm}^2)$$
$$\Sigma_H(\text{cm}^{-1}) = n_{c\ H}(\text{cm}^{-3}) \cdot \sigma_H(\text{cm}^2) = (6.690 \cdot 10^{22}) \times (4.52 \cdot 10^{-24}) = 0.30 \text{ cm}^{-1}$$
$$\Sigma_O(\text{cm}^{-1}) = n_{c\ O}(\text{cm}^{-3}) \cdot \sigma_O(\text{cm}^2) = (3.345 \cdot 10^{22}) \times (3 \cdot 10^{-24}) = 0.10 \text{ cm}^{-1}$$

therefore:

$$\Sigma_{tot} = \Sigma_H + \Sigma_O = 0.4 \text{ cm}^{-1}$$

3. Calculate the fluence ϕ for the radiation emerging from the water shield.

$$\phi = \phi_0 e^{-\Sigma_{tot} x}$$

therefore:

$$\phi = 1\,000\,000 e^{-0.4 \times 10} = 18\,315 \text{ neutrons per cm}^3$$

2.5. Check your knowledge

You may now test what you have learned in studying this chapter, by finding the answers to the following questions:

1. **List the modes of electron interactions with matter. Of these various phenomena, are some found to be dominant? If so, what are the factors promoting this?**

    Answer:

    Ionisation, excitation, bremsstrahlung + matter–antimatter annihilation for positrons; ionisation and excitation are dominant compared to bremsstrahlung; the latter can become significant for very-high-energy (several MeV) electrons passing through dense, high-Z materials.

2. **Explain the various interaction modes reviewed in Question 1, by means of annotated diagrams.**

   Answer:

   See section 2.2.

3. **An electron endowed with a starting energy of 900 keV has an average linear energy transfer (LET), in soft tissue, of 2.6 MeV·cm^{-1}. What is its pathlength in such tissue?**

   Answer:

   $p = 900/2600 = 0.35$ cm

4. **An electron with a starting energy of 80 keV has an average linear energy transfer (LET), in soft tissue, of 4.75 MeV·cm^{-1}. What pathlength will it cover?**

   Answer:

   $p = 80/4750 = 0.17$ mm

5. **An electron with a starting energy of 1.2 MeV traverses a pathlength of 4.8 mm in soft tissue. Calculate its average linear energy transfer (LET).**

   Answer:

   LET = $1.2/0.48 = 2.5$ MeV·cm^{-1}

6. **An electron with a starting energy of 500 keV covers a pathlength of 0.25 cm in soft tissue. Calculate its average linear energy transfer (LET).**

   Answer:

   LET = $500/0.25 = 2000$ keV·cm^{-1} = 2 MeV·cm^{-1}

7. **The linear energy transfer (LET) of an electron endowed with an energy of 250 keV is found to stand at 2.5 MeV·cm^{-1} in soft tissue. Calculate the quantity of energy that is transferred to tissue as this electron traverses a pathlength of one tenth of a millimeter.**

   Answer:

   2500 keV·cm^{-1} = 250 keV·mm^{-1} = 25 keV per 1/10 mm (i.e. per 0.1 mm)

8. **The linear energy transfer (LET) of an electron with an energy of 500 keV is equal to 2 MeV·cm^{-1} in soft tissue. Calculate the quantity of energy transferred to tissue as this electron covers 0.25 mm.**

   Answer:

   2000 keV·cm^{-1} = 200 keV·mm^{-1} = 50 keV per 1/4 mm (i.e. per 0.25 mm).

9. **Linear energy transfer (LET) varies according to the energy that directly ionising particles are endowed with. In the case of electron irradiation, at low energies, does LET increase, or decrease as energy rises? What may we infer from this in the event of exposure of living tissue to electron irradiation?**

   Answer:

   At low energies, LET decreases as energy rises. It follows, as regards living tissue, that the lower the energy of an electron, the greater its ability becomes of causing localized damage.

10. **By what phenomenon is beta-plus radiation accompanied, in matter?**

    Answer:

    Positron annihilation, once the particle has lost all of its kinetic energy. Each annihilation event yields two gamma-rays, each with an energy of 511 keV, emitted in opposite directions (i.e. at an angle of 180°).

11. **What is the order of magnitude for the range of alpha radiation in air?**

    a 5 mm

    b 5 cm

    c 50 cm

    d 5 m

    Answer:

    Alpha radiation ranges out to 5 cm or so in air.

12. **What is the order of magnitude for the range of beta radiation in air?**

    a several micrometers

    b several millimeters

    c several meters

    d several hundred meters

    Answer:

    The range for such radiation is several meters in air.

13. **What is the dominant interaction process, in matter, for gamma radiation with an energy of around 1 MeV?**

    a the photoelectric effect

    b the Compton effect

    c pair production

    d bremsstrahlung

    Answer:

    The Compton effect.

14. **What is the dominant interaction process, in matter, for low-energy gamma radiation?**

    a the photoelectric effect

    b the Compton effect

    c pair production

    d bremsstrahlung

    Answer:

    The photoelectric effect.

15. Which is the interaction process of gamma radiation with matter that can only occur for energies higher than a theoretical threshold of 1.02 MeV?

a the photoelectric effect

b the Compton effect

c pair production

d bremsstrahlung

Answer:

Pair production, also known as the materialization effect.

16. Why are gamma or X-rays known as indirectly ionising radiation?

Answer:

Because gamma radiation generates, or sets in motion electrons, which in turn ionize the material with which that radiation has interacted.

17. Calculate the range in air of β radiation found to have an average energy of 100 keV. The following formula is given:

$R = \dfrac{0.412 \, E^n}{\rho}$ with: $n = 1.265 - 0.0954 \ln E$

where E is the maximum electron energy, ρ the density of air (taken to be equal to $1.3 \cdot 10^{-3}$ g·cm^{-3}).

Answer:

First, the electrons' maximum energy must be determined:

This being β radiation, the maximum energy for the β spectrum is about 3 times the value of the average energy, i.e.: $E_{max} = 3 E_{av} = 300$ keV $= 0.3$ MeV.

Next, the value for exponent n can be calculated:

$n = 1.265 - 0.0954 \ln(0.3) = 1.380$

In air, since density has a value of $1.3 \cdot 10^{-3}$ g·cm^{-3}, performing the calculation as indicated by the formula provided yields the range:

$R_{air} = (0.412 \times 0.3^{1.38})/(1.3 \cdot 10^{-3}) = 60$ cm

18. Complete the text in the following table by filling in the gray areas:

Table 2.3. The chief interactions of ionising radiation with matter, and secondary emitted radiation (corrected question version)

Radiation	Type of interaction	Radiation emitted after interaction
▓▓▓ particles (α, electrons)	▓▓▓	– an ▓▓▓ in the material traversed is set in motion – rearrangement of the electron cloud (X-rays, and/or Auger electrons)
	▓▓▓	– rearrangement of the electron cloud (X-rays, and/or Auger electrons)
	Bremsstrahlung (only for ▓▓▓)	– X-ray emission (bremsstrahlung)
	▓▓▓ (end of path for ▓▓▓)	– emission of ▓▓▓ photons of ▓▓▓ keV at 180°
▓▓▓ (γ, X)	Photoelectric effect (▓▓▓)	– ▓▓▓ of an atomic electron – rearrangement of the electron cloud (X-rays, and/or Auger electrons)
	▓▓▓ (intermediate E)	– ▓▓▓ of an atomic electron – scattering of the incident photon – rearrangement of the electron cloud (X-rays, and/or Auger electrons)
	Pair production ▓▓▓	– ▓▓▓
Neutrons	▓▓▓	– emission of a secondary ▓▓▓ (neutron, proton, α, γ …)
	Scattering	– scattering of the incident ▓▓▓ – propulsion of recoil nucleus

Answer:

Table 2.3. The chief interactions of ionising radiation with matter, and secondary emitted radiation (corrected answer version)

Radiation	Type of interaction	Radiation emitted after interaction
Charged particles (α, electrons)	Ionisation	– an atomic electron in the material traversed is set in motion – rearrangement of the electron cloud (X-rays, and/or Auger electrons)
	Excitation	– rearrangement of the electron cloud (X-rays, and/or Auger electrons)
	Bremsstrahlung (only for electrons)	– X-ray emission (bremsstrahlung)
	Annihilation (end of path for positrons)	– emission of 2 × 511-keV photons at 180°
Electromagnetic radiation (γ, X)	Photoelectric effect (low energies)	– ejection of an atomic electron – rearrangement of the electron cloud (X-rays, and/or Auger electrons)
	Compton effect (intermediate E)	– ejection of an atomic electron – scattering of the incident photon – rearrangement of the electron cloud (X-rays, and/or Auger electrons)
	Pair production ($E > 1.022$ MeV)	– generation of a negatron–positron pair
Neutrons	Absorption	– emission of a secondary radiation (neutron, proton, α, γ …)
	Scattering	– scattering of the incident neutron – propulsion of recoil nucleus

Further information

19. Emission from a source results in a fluence of $2 \cdot 10^6 \, \gamma \cdot \text{cm}^2$ It is secured in a container with walls consisting of 11 cm aluminum, and 3 cm lead. Calculate the fluence that now obtains. Data:

    for aluminum: $\mu/\rho = 0.0613 \text{ cm}^2 \cdot \text{g}^{-1}$; $\rho = 2.7 \text{ g} \cdot \text{cm}^{-3}$

    for lead: $\mu/\rho = 0.0708 \text{ cm}^2 \cdot \text{g}^{-1}$; $\rho = 11.34 \text{ g} \cdot \text{cm}^{-3}$

    Answer:
    Value for the fluence, after traversing 11 cm aluminum:

    $$\phi = \phi_0 e^{-\frac{\mu}{\rho}\rho x} = 2 \cdot 10^6 \times e^{-0.0613 \times 2.7 \times 11} = 3.24 \cdot 10^5 \, \gamma \cdot \text{cm}^{-2}$$

    Value for the fluence, after traversing 3 cm lead:

    $$\phi = \phi_0 e^{-\frac{\mu}{\rho}\rho x} = 3.24 \cdot 10^5 \times e^{-0.0708 \times 11.34 \times 3} = 2.9 \cdot 10^4 \, \gamma \cdot \text{cm}^{-2}$$

3

Dosimetry

Hervé Viguier, Alain Vivier

Introduction

Ionising radiation necessarily affects living organisms – or indeed any material – exposed to it. To assess the damage occasioned by such exposure, the dose relating to a given organ, or material, must first be ascertained. Dosimetry is the tool that serves to measure doses due to ionising radiation, using methods determined in accordance with radiation protection concepts and criteria, as formulated, in particular, by the International Commission on Radiological Protection (ICRP).

This chapter sets out to review the concepts used in dosimetry, so as to give an understanding of the **physical quantities** involved. To every physical quantity encountered, there corresponds a particular concept, and a specific unit, which need to be properly understood, to preclude any risk of confusion. Our purpose will be to show how these units are to be interpreted, and how they relate to one another, in a manner that, on the basis of a given number of particles, allows the energy deposited in any given mass – whether living, or inert – to be determined. We will then consider the radiological protection quantities defined by the ICRP, created to estimate the risk and to comply with regulatory limits. As these body-related protection quantities are not measurable, the concept of operational quantities has been introduced to give access to reasonable assessments of the radiation protection quantities. So the presentation of these quantities will close this chapter. In conclusion, the reading of this chapter will give a complete overview of all the quantities related to the concept of "dose".

For practical calculations of radiation protection, the radiometric quantities used in this chapter are intentionally simplified. We restrict ourselves to calculations of macroscopic quantities.

3.1. Physical quantities

The dose is a quantity that serves to quantify the effects occasioned by radiations, in a person exposed to them. A dose may be estimated on the basis of the irradiation conditions, when these are known: nature of the radionuclides involved, activity, relative position... Or it may be more straightforwardly measured, by means of personal dosimeters, or dosimeters monitoring ambient conditions. Put simply, two main types of dose may be distinguished:

- the "**physical dose**", measured in grays (Gy), which is simply the quantity of energy deposited per unit mass: this corresponds to the absorbed dose, which can be directly measured;

- the "**radiological protection dose**", expressed in sieverts (Sv), which covers both the equivalent dose, and the effective dose, both calculated on the basis of the absorbed dose, which serve to quantify biological effects, with regard to stochastic – and consequently long-term – effects, for low dose levels (risks of cancer onset).

3.1.1. Absorbed dose

The **absorbed dose** is the basic physical quantity; it is used for all types of ionising radiation, and any irradiation geometry. It is defined as the quotient of dE divided by dm, where dE is the mean quantity of energy imparted to matter of mass dm by the ionising radiation of interest, viz.:

$$D = \frac{dE}{dm}$$

The unit of absorbed dose, in the International System of Units (SI), is the joule per kilogram ($J \cdot kg^{-1}$); its special name is the gray (Gy). The definition of the absorbed dose takes no account of the nature of the matter affected, or of the nature and energy of the radiations involved. It is a strictly physical quantity.

A dose of 1 Gy = 1 $J \cdot kg^{-1}$ is a very high dose; in practice, submultiples of the gray are used, e.g. the milligray (mGy), or the microgray (μGy). For living matter, the gray is the unit of dose used to quantify high dose levels. The dose received from a chest (pulmonary) X-ray is of the order of 1 milligray. In radiotherapy, the doses imparted within tumors range up to 50 Gy. Such doses, which result in cell deaths on a massive scale, are nonetheless tolerated by the organism, as they are highly localized. On the other hand, a "whole-body" dose of 10 Gy results in death in most cases (lethal dose). The 50%-lethality dose (i.e. the dose that results in death for one half of the individuals receiving it, within 60 days) stands at 4.5 Gy.

In the event of irradiation, part of the energy is absorbed by the ambient atmosphere, and by the materials encountered by the radiation before it reaches the sample, or the individual. For the purposes of dose calculations, such geometrical and absorption factors must be taken into account. In principle, the absorbed dose is obtained by multiplying the various coefficients of interest by the source's activity, and by the mean energy deposited by the radiation involved, per incident particle. Such calculations are fairly complex, and call for the use of special computation codes. In practice, dosimeters are employed, to arrive at an empirical dose evaluation.

The absorbed-dose rate, noted $\dot{D}$, indicates the rate at which the dose is delivered. In the International System of Units, the absorbed-dose rate is measured in units of grays per second ($Gy \cdot s^{-1}$); however, owing to the low levels of activity encountered, submultiples are commonly used, e.g. the milligray per hour ($mGy \cdot h^{-1}$), or the microgray per hour ($\mu Gy \cdot h^{-1}$).

If the absorbed dose-rate remains constant over the time interval t considered, the following relation holds:

$$D = \dot{D} t$$

Example:

If the absorbed-dose rate due to the work environment at a workstation stands at 0.3 mGy·h^{-1}, and if the operator remains at his post for 2 hours 20 minutes, the dose absorbed by that worker's body, as a whole, is equal to:

$$D = 0.3 \times \left(2 + \frac{20}{60}\right) = 0.7 \text{ mGy}$$

3.1.2. Relation between dose and fluence

When radiations (be they charged particles, photons, or neutrons) penetrate a medium, they have the ability, by way of interaction mechanisms acting directly or indirectly, to transfer all or part of their initial (starting) energy E to matter, this resulting in the absorption of part or all of this transferred energy in the material involved. The energy balance, in principle, for this absorbed dose may be determined by a straightforward physical calculation.

Consider N incident particles each endowed with kinetic energy E, forming a beam which penetrates a material of mass m. All or part of the kinetic energy of every particle is thus absorbed by the material. If the mean energy deposited per particle is noted E', the absorbed dose may now be written as follows:

$$D = \frac{E_{\text{abs}}}{dm} = \frac{E' N}{m}$$

Bearing in mind that: $m = \rho A x$, where ρ is the material's density, A the surface area of the body of material, x the material thickness, this becomes:

$$D = \frac{E' N}{\rho A x} = \left(\frac{E'}{\rho x}\right)\left(\frac{N}{A}\right) = f \Phi$$

where $\Phi = \frac{N}{A}$ stands for the number of incident particles, per unit surface area. This quantity, known as the **fluence**, is expressed, as a rule, in terms of a number per cm^2 (i.e. in cm^{-2}).

The quantity $E'/(\rho x)$ is identified as a coefficient f, referred to as the **fluence-to-dose conversion coefficient**, which is dependent only on the nature, and energy E of the incident radiation, and on the constituent elements of the target material. This coefficient f is expressed in units of MeV·cm^2·g^{-1}. For any given source, this coefficient is commonly found to be a constant. It follows that the absorbed dose is then proportional to fluence.

It may be found advantageous, in some cases, to determine the fluence per unit time, i.e. the fluence rate, noted $\dot{\Phi}$, which serves to calculate the absorbed-dose rate.

3.1.3. Dose calculations for charged particles

As they pass through matter, charged particles interact with the atoms or molecules lying along their path. These interactions often take place in gradual fashion (ionisations, excitations), although they may occur more abruptly for low-mass particles, e.g. electrons (bremsstrahlung). As a result, these particles are seen to sustain a gradual, continuous energy loss, over their entire path. In such conditions, a quantity known as the **mean linear stopping power**, noted $\bar{S}$, may be defined, equal to the mean energy loss, for a particle, per unit length of path traversed:

$$\bar{S} = \frac{dE}{dx}$$

In practice, the linear stopping power is expressed in units of keV·μm^{-1} in dense materials, or in keV·cm^{-1} in water. The stopping power depends on the nature, and energy of the charged particles involved, and on the nature, in terms of chemical elements, of the material traversed. It is further found to be proportional to the material's density. The Bethe formula, or experimental measurements may serve to determine its value.

For the purposes of dose calculations, the mass stopping power must be determined: this is obtained as the quotient of the linear stopping power by the material's density:

$$\frac{\bar{S}}{\rho} = \bar{S}\frac{1}{\rho} = \frac{dE}{dx\rho}$$

It can be shown that, for charged particles, the fluence-to-dose conversion coefficient f is equal to the **mass stopping power**. Consequently, the dose calculation, making use of the mass stopping power, may be written as:

$$D = f\Phi = \frac{\bar{S}}{\rho}\Phi$$

If the mass stopping power and the fluence are known, it thus becomes possible to determine the mean absorbed dose in a medium.

The mass stopping power rises as the particle's velocity diminishes. In effect, the more slowly a particle travels, the more time it has to interact with the medium, thus depositing part of its energy as it goes. The relation between mass stopping power and range is shown as a plot known as the **Bragg curve** (an instance of which is shown as Figure 3.1). This

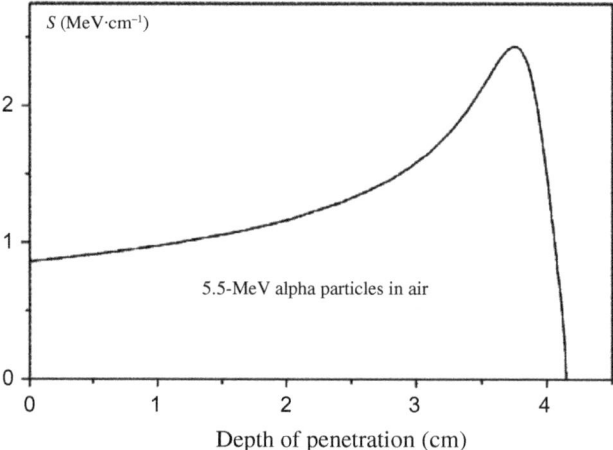

Figure 3.1. The Bragg curve.

features a peak at the end of the path, indicating a steep rise in mass stopping power as the particle nears the end of its path.

As a rough-and-ready approximation, to calculate the dose, the mass stopping power may be regarded as remaining constant over the length of the particle's path. Its variation with particle velocity is thus disregarded, once it has entered the medium, and an average mass stopping power value is considered. Such values are listed in data tables, or can be read off reference curves.

3.1.3.1. *Dose calculations for heavy charged particles*

Protons, alpha particles, and other such nuclei acting as projectiles are referred to as heavy particles, compared to the electrons they impinge on. Indeed, protons weigh in at some 2000 electron masses (proton rest mass-energy: $m_p c^2 = 938$ MeV; electron rest mass-energy: $m_e c^2 = 0.511$ MeV). α particles, in turn, are some 8000 times heavier than electrons.

Such large masses result in three major consequences, as far as we are concerned here.

Whenever a heavy particle collides with an electron, as it travels through the medium, no deflection of the incident particle ensues. It follows directly from this that such particles, in matter, travel along straight paths (except for some exceedingly rare collisions with a nucleus).

The velocity of these particles is lower, for a given kinetic energy, than that of an electron of the same energy. As a result, heavy particles "spend more time" close to the electrons in the medium, this occasioning much higher exchanges of energy. Heavy particles thus all involve much higher stopping powers, compared to electrons. For a given kinetic energy, the absorbed dose – which is proportional to the stopping power – is higher than that found for electrons.

The energy transferred to the medium is equal to the energy absorbed by that medium, as the sole transfer mechanism involved, for heavy charged particles, is the collision process, i.e. essentially ionisations, and excitations. This is not the case for incident electrons, owing to bremsstrahlung interactions, which result in less energy remaining to be absorbed by the medium. If the material thickness is greater than the particle range R, the quantity of energy absorbed, which serves to calculate the absorbed dose, stands equal to the particle's starting energy. This finding allows the mean linear stopping power, for a heavy particle, to be readily evaluated, by means of the following relation:

$$\bar{S} = \frac{E_i}{R}$$

where E_i is the heavy particle's incident energy, R the particle's range in the material.

By way of comparison, Table 3.1 shows the mean linear stopping powers, in water, for alpha particles and protons. To calculate the absorbed dose, these values just need to be divided by the material's density, and multiplied by the fluence.

It should be borne in mind that the value so calculated, for absorbed dose, is valid across a thickness R of material. Further out, the dose comes down to zero.

Finally, in order to arrive at the mean dose, the relation between heavy-particle fluence and mean stopping power must be brought in:

$$D = \frac{\bar{S}}{\rho} \Phi$$

Table 3.1. Ranges and mean linear stopping powers for some heavy charged particles.

Energy E_i (keV)	Nature of particle	Range (water) R (μm)	Stopping power $\bar{S}$ (keV/μm)
10	alpha	0.27	37
	proton	0.35	29
100	alpha	1.4	71
	proton	1.6	63
500	alpha	3.69	136
	proton	8.9	56
1000	alpha	5.93	169
	proton	24.5	41
3000	alpha	18.2	165
	proton	150	20
5000	alpha	37.5	133
	proton	362	14
10 000	alpha	113	88
	proton	1230	8

In practice, the units used are not SI units, and the above relation then reads:

$$D = 1.6 \cdot 10^{-7} \frac{\bar{S}}{\rho} \Phi$$

where the mean absorbed dose D is expressed in mGy, the mean mass stopping power $\frac{\bar{S}}{\rho}$ in MeV·cm²·g⁻¹, fluence Φ in particles per cm² (i.e. in cm⁻²).

If the fluence rate is used, rather than the fluence, the following relation is obtained:

$$\dot{D} = 5.76 \cdot 10^{-4} \frac{\bar{S}}{\rho} \dot{\Phi}$$

where the mean absorbed-dose rate $\dot{D}$ is expressed in mGy·h⁻¹, the mean mass stopping power $\frac{\bar{S}}{\rho}$ in MeV·cm²·g⁻¹, the fluence rate $\dot{\Phi}$ in particles·cm⁻²·s⁻¹.

Exercise:

A source emits 5000-keV alpha particles. The particle beam is found to result in a fluence rate of 10^3 α particles per cm² per second.

1. *Calculate the absorbed dose in water after 20 minutes.*

When the fluence rate is integrated over 20 minutes, the fluence is found to be equal to: $\Phi = 20 \times 60 \times 10^3 = 1.2 \cdot 10^6$ α particles·cm⁻². Table 3.1 gives a value: $\bar{S} = 133$ keV·μm⁻¹, i.e. $\bar{S} = 1330$ MeV·cm⁻¹. Taking $\rho_{water} = 1$ g·cm⁻³, the mean mass stopping power, for 5000-keV α particles, in water, is found to be equal to: $\frac{\bar{S}}{\rho} = 1330$ MeV·cm²·g⁻¹. Therefore:

$$D = 1.6 \cdot 10^{-7} \frac{\bar{S}}{\rho} \Phi = 1.6 \cdot 10^{-7} \times 1330 \times 1.2 \cdot 10^6 = 255 \text{ mGy}$$

Note: this is an extremely high dose. On the other hand, as regards external exposure, for an individual, such a dose would be of no consequence, in biological terms. Indeed, it will be seen, from the same table, that this dose is deposited over a thickness of 37 μm. Now the dead skin layer, as a rule, is found to be 70 μm thick, thus providing full protection against such external radiation.

2. *Calculate the absorbed-dose rate in water.*

The formula involving the fluence rate is used:

$$\dot{D} = 5.76 \cdot 10^{-4} \frac{\bar{S}}{\rho} \dot{\Phi} = 5.76 \cdot 10^{-4} \times 1330 \times 10^3 = 766 \text{ mGy} \cdot \text{h}^{-1}$$

3.1.3.2. Dose calculations for electrons

Being charged particles as they are, electrons give rise to interactions with matter that present great similarities with those arising with heavy charged particles.

Electrons that penetrate a medium likewise undergo a continuous slowing down, due to successive collisions with electrons in the material they are traveling through, the energy lost in this manner corresponding to the energy transferred to the material. Differences are also apparent, however, owing to the mass of the electron.

Collisions with atomic electrons – which are thereby set in motion, as so-called "secondary electrons" – since these have the same mass as the incident ("primary") electron, are liable to result in considerable deflections in the latter's path. The primary electron's path thus exhibits a tortuous aspect, which becomes more marked as the electron's energy diminishes. It will be remembered that, for heavy charged particles, paths are effectively straight.

The velocity of electrons, for a given kinetic energy, is higher than would be the case for heavy charged particles (electron velocities approach the speed of light, for energies higher than 50 keV). Consequently, stopping powers, in any given material, are much lower, for electrons, than the stopping powers found for heavy charged particles. Conversely, for a given starting kinetic energy, an electron's range across a material is found to be much longer (of the order of a few millimeters) than is the case for heavy charged particles (a few micrometers). By way of comparison, Table 3.2 shows the mean linear stopping powers, in water, for alpha particles and electrons.

Electrons interact both by way of collisions (ionisations, excitations), and through bremsstrahlung. The latter effect is altogether negligible for heavy charged particles. Bremsstrahlung ("braking radiation") carries an energy corresponding to the kinetic energy lost by the electron when it is deflected as it passes close to an atomic nucleus in the medium it is going through. This energy loss results in emission of a bremsstrahlung X-photon, which, as a rule, passes out of the local interaction site, without contributing to energy absorption in the material. The fraction of energy that is so emitted as bremsstrahlung becomes all the higher, with increasing electron kinetic energy; and higher material density (higher Z).

The mean linear stopping power, for electrons, is equal to the sum of the **collisional stopping power** $\bar{S}_{col}$, and the **radiative stopping power** $\bar{S}_{rad}$, i.e. the component due to bremsstrahlung:

$$\bar{S}_{tot} = \bar{S}_{col} + \bar{S}_{rad}$$

Table 3.2. Comparison of ranges and mean linear stopping powers for electrons and alpha particles.

Energy E_i (keV)	Nature of particle	Range (water) R (μm)	Stopping power $\bar{S}$ (keV/μm)
100	alpha	1.4	71
	electron	140	0.71
500	alpha	3.69	136
	electron	1700	0.29
1000	alpha	5.93	169
	electron	4300	0.23
5000	alpha	37.5	133
	electron	25 500	0.20
10 000	alpha	113	88
	electron	49 700	0.20

For the purposes of calculating the dose, only the collisional stopping power need be considered. If the value for the total stopping power is known, the value for the radiative stopping power must be subtracted from it. The following approximate formula serves to determine the radiation yield, rated g:

$$g = \frac{\bar{S}_{rad}}{\bar{S}_{tot}}$$

The collisional stopping power is then derived by way of the following relation:

$$\bar{S}_{col} = \bar{S}_{tot}(1 - g)$$

We can also define the relationship between the radiative stopping power and the collisional stopping power according to the relationship:

$$\frac{\bar{S}_{rad}}{\bar{S}_{col}} \approx \frac{ZE}{715}$$

where Z is the material's atomic number, E the electrons' incident energy (in MeV).

It can be seen that the radiation yield becomes negligible for low electron energies, and in low-Z materials.

Owing to the far more erratic paths followed by electrons, calculating the absorbed dose proves to be a more complex task. However, the numerous scattering events electrons are subject to do tend to result in a more homogeneous dose distribution, within a thickness of material corresponding to their range. The mean dose thus stands as a good approximation for the actual dose, at every point along the electrons' path, particularly since the stopping power varies far less markedly, over the entire path, than is the case for heavy charged particles.

By way of example, we can look at the energy loss, and the stopping power, in water, for a 1-MeV electron, as a function of the thickness traversed (see Figure 3.2).

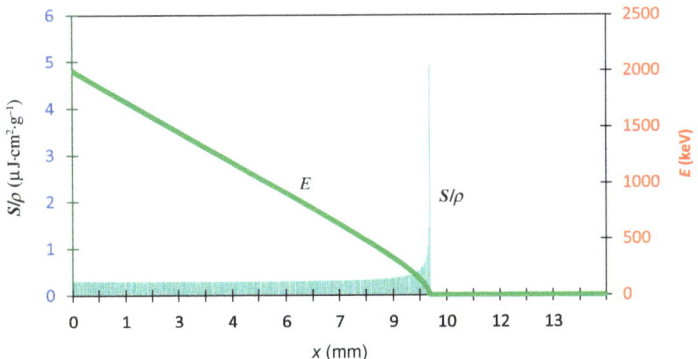

Figure 3.2. Right-hand scale: energy loss for an electron, in water. Left-hand scale: mass stopping power S/ρ for an electron, in water.

Bearing all this in mind, the mean stopping power, over the entire range, in a sufficiently thick target, may thus be taken to be equal to:

$$\bar{S}_{tot} = \frac{E_i}{R}$$

and:

$$\bar{S}_{col} = \bar{S}_{tot}\left(1 - \frac{ZE}{715}\right)$$

Then, as in the case of heavy charged particles, the relation between charged-particle fluence, and mean collisional mass stopping power must be set out:

$$D = \frac{\bar{S}_{col}}{\rho}\Phi$$

In practice, the units used are not SI units, and the above relation then reads:

$$D = 1.6 \cdot 10^{-7} \frac{\bar{S}_{col}}{\rho}\Phi$$

where the mean absorbed dose D is expressed in mGy, the mean collisional mass stopping power $\frac{\bar{S}_{col}}{\rho}$ in MeV·cm²·g⁻¹, fluence Φ in particles per cm² (i.e. in cm⁻²).

If the fluence rate is used, rather than the fluence, the following relation is obtained:

$$\dot{D} = 5.76 \cdot 10^{-4} \frac{\bar{S}_{col}}{\rho}\dot{\Phi}$$

where the mean absorbed-dose rate $\dot{D}$ is expressed in mGy·h⁻¹, the mean collisional mass stopping power $\frac{\bar{S}_{col}}{\rho}$ in MeV·cm²·g⁻¹, the fluence rate $\dot{\Phi}$ in particles·cm⁻²·s⁻¹.

Electrons from the beta emission spectrum of a radionuclide are endowed with energies that form a continuous distribution, ranging from 0 to $E_{\beta\,max}$. This energy spectrum is characteristic of the particular beta-emitter radionuclide. The dose calculation then becomes quite complex, since the electrons may have a wide variety of ranges, resulting in diverse values for the stopping power. A fairly rough-and-ready approach involves calculating the dose by taking the average beta energy for the spectrum of interest. The above formulae are then used as though all the beta particles were endowed with that same average energy.

3.1.4. Dose calculations for γ- and X-photons

The photons in gamma rays, and X-rays are uncharged particles: in contrast to charged particles, the interactions they give rise to, in matter, occur in random fashion. A photon may reemerge without having undergone any interaction; nor is there any gradual loss of energy, transferred to the medium. The concept of stopping power thus cannot be used to evaluate the energy lost by photons inside the material. On the other hand, there is a probability of interaction, which varies depending on photon energy, and the nature of the material traversed. When an interaction does occur – with a single electron – it is characterized by an exchange of energy between the photon, and the electron. The latter then carries off all of the energy gained in the energy transfer, acting as a primary electron, endowed with a varying amount of energy. It is this electron that in turn deposits, across the medium, the energy transferred by the photon, thus contributing to the absorbed dose.

In order to calculate the absorbed dose, the number of photons must be ascertained, that interact with the medium, at a given depth. The following attenuation law will be remembered, from Chapter 2, "Interaction of ionising radiation with matter", characterizing as it does the probabilistic phenomenon of photon interactions in matter:

$$N_{emergent} = N_0 e^{-\mu x}$$

where $N_{emergent}$ is the number of photons that emerge from a thickness x of material without interacting, N_0 is the number of photons entering the material, μ is the attenuation coefficient appropriate for the photon, in that material, x is the thickness of material.

If we consider a thin target, of thickness dx, the calculation of the number of photons that interact inside the material may be written as follows:

$$N_{interacting} = N_0 - N_{emergent} = N_0 - N_0 e^{-\mu dx}$$

As $\mu dx \ll 1$, we may write: $e^{-\mu dx} = 1 - \mu dx$.

The above relation may thus be simplified as follows:

$$N_{interacting} = N_0 - N_0 e^{-\mu dx} = N_0[1 - (1 - \mu dx)] = N_0 \mu dx$$

It follows that the number of photons that interact in the material is proportional to the number of incoming photons, to the small thickness of material, and to the attenuation coefficient.

In order to determine the absorbed dose, we must first determine the quantity of energy that is transferred to the medium, for every photon that interacts. We have seen that photon interactions – involving the photoelectric effect, the Compton effect, or pair production – are accompanied by γ- or X-ray scattering or emission phenomena. As a result, a certain fraction of the incident photon's energy is able either to interact in turn with the medium, or escape out of the material.

Over a large number of interactions in a material, involving photons of energy E_γ, the mean quantity of energy $\bar{E}_{tr}$ transferred to an electron, in every interaction, is equal to a fraction k of E_γ, so that:

$$\bar{E}_{tr} = kE_\gamma$$

A linear energy-transfer coefficient may thus be defined – as a rule simply referred to as the **energy-transfer coefficient** – by way of the following relation:

$$\mu_{tr} = k\mu$$

The mean transferred energy may then be written as:

$$\bar{E}_{tr} = \frac{\mu_{tr}}{\mu} E_\gamma$$

The quantity of energy transferred to the material thus stands equal to the product of the number of photons that have undergone an interaction, by the mean transferred energy, for each interacting photon:

$$E_{transferred} = N_{interacting}\bar{E}_{tr} = (N_0 \mu\, dx)\left(\frac{\mu_{tr}}{\mu} E_\gamma\right) = N_0 \mu_{tr} E_\gamma\, dx$$

3.1.4.1. The kerma concept

The kerma (kinetic energy released per unit mass) is the quantity of energy transferred per unit mass of material. This quantity has the same dimensions as absorbed dose, and is likewise expressed in grays. It is not to be identified with the absorbed dose, however. Indeed, the kerma is defined as the energy transferred at a given point in the material – at the site of interaction – whereas the absorbed dose arises, due the electron thus set in motion, some distance away from the photon interaction site. Moreover, not all of the transferred energy gets to be absorbed, since a fraction of the electron's energy ends up in the form of bremsstrahlung (this phenomenon is negligible in biological tissue).

The kerma may thus be written as:

$$K = \frac{E_{transferred}}{m} = \frac{N_0 \mu_{tr} E_\gamma\, dx}{m}$$

Bearing in mind that: $m = \rho A dx$, where ρ is the material's density, A the surface area of the body of material, dx the material thickness, this becomes:

$$K = \frac{N_0 \mu_{tr} E_\gamma\, dx}{\rho A dx} = \frac{\mu_{tr}}{\rho} E_\gamma \frac{N_0}{A}$$

Substituting $N_0/A = \Phi$, this expression simplifies to:

$$K = \frac{\mu_{tr}}{\rho} E_\gamma \Phi$$

The quantity $\frac{\mu_{tr}}{\rho} \cdot E_\gamma$ thus stands as the fluence-to-kerma conversion coefficient.

The quantity $\frac{\mu_{tr}}{\rho}$ is the mass absorption coefficient ($cm^2.g^{-1}$).

If the kerma is to be calculated for a thick target material, the above relation now reads:

$$K(x) = \frac{\mu_{tr}}{\rho} E_\gamma \Phi e^{-\mu x}$$

where x is the material thickness, μ the total attenuation factor.

The kerma stands as a crucial quantity, with regard to radiation measurement. It corresponds to **the dose transferred by the electromagnetic radiation, not to the absorbed dose however**. The latter may stand equal to the transferred dose, provided electronic equilibrium is attained (see Box "Further information" below), and bremsstrahlung is negligible.

Further information

The concept of electronic equilibrium

When interactions occur, the quantity of energy transferred to electrons is high. The electrons involved then traverse distances of varying magnitude, usually of the order of a few millimeters, or a few centimeters in low-Z materials (water, biological tissue…). As they are slowed down, they deposit their energy, distributing it along their entire path. The energy absorbed by the medium thus does not remain local, rather it is distributed over the entire electron pathlength. The absorbed dose thus depends on the fluence of secondary electrons at every point in the medium.

To understand the notion of equilibrium, let us suppose the material is divided into thin segments, and that an equal amount of photon interaction takes place in every one of these segment volumes (see example in Figure 3.3). In each volume, the number of electrons liberated and set in motion is added to the number of electrons liberated in the volumes before it – and so on from one volume to the next. The electron fluence thus builds up as depth increases, as does the absorbed dose – through to a depth equal to the maximum electron range. At that depth, the number of electrons set in motion that enter each volume becomes equal to the number of electrons set in motion that leave it. At this point, electronic equilibrium is said to exist. In this state – aside from bremsstrahlung – the quantity of energy transferred stands equal to the quantity of energy absorbed locally.

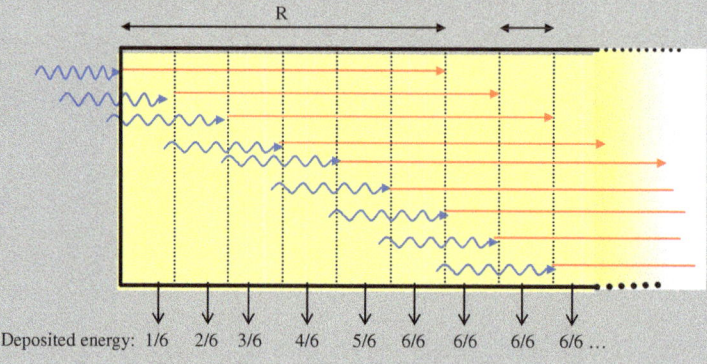

Figure 3.3. Principle schematic of electronic equilibrium.

> In fact, the electron fluence depends on the photon fluence, which may be taken to remain effectively constant across the first few segments, but decreases in accordance with the photon attenuation law over greater depths. The outcome is heavily dependent, of course, on the electromagnetic radiation's energy.
>
> To sum up: electronic equilibrium is attained when the dose is measured at a depth greater than the range of the primary electrons.

3.1.4.2. Absorbed dose from electromagnetic radiation

Once electronic equilibrium is attained, if no account is taken of bremsstrahlung, the absorbed dose may be determined according to the following relation:

$$D(x) = K(x) = \frac{\mu_{tr}}{\rho} E_\gamma \Phi e^{-\mu x}$$

In such conditions, absorbed dose and kerma are equal, since the quantity of energy absorbed locally, at every point, stands equal to the quantity of energy transferred to the material.

In cases where radiative losses – chiefly bremsstrahlung – cannot be so disregarded, the absorbed dose calculation must take into account the mean fraction of energy transferred to electrons that is lost in radiative processes, noted g. Conversely, the fraction corresponding to the mean quantity of kinetic energy not carried off by bremsstrahlung or other radiative processes stands equal to $1 - g$. The calculation now entails reducing the above expression by this factor $1 - g$:

$$D(x) = (1-g)\frac{\mu_{tr}}{\rho} E_\gamma \Phi e^{-\mu x}$$

Absorbed dose and kerma are no longer equal, since the quantity of energy that is absorbed locally is now smaller than the quantity of energy transferred to the medium.

A linear energy-absorption coefficient, noted $\mu_{en} = (1-g)\mu_{tr}$, may thus be defined – simply referred to as the **energy-absorption coefficient** – and the following expression is obtained, valid for electron equilibrium:

$$D(x) = \frac{\mu_{en}}{\mu_{tr}}\frac{\mu_{tr}}{\rho} E_\gamma \Phi e^{-\mu x}$$

Which yields the final expression for the absorbed dose, once electronic equilibrium is attained:

$$D(x) = \frac{\mu_{en}}{\rho} E_\gamma \Phi e^{-\mu x}$$

Note: in biological tissue, the mean fraction of energy lost by bremsstahlung is quite low – for instance: $g = 0.002$ for 1-MeV photons.

The fluence-to-dose conversion coefficient f thus reappears, once electronic equilibrium is attained, then being equal to $\frac{\mu_{en}}{\rho} E_\gamma$.

Table 3.3. **Mass energy-absorption coefficients for photons in water, or tissue.**

E (keV)	10	50	100	200	300	500	1000	1500	3000	5000
$\frac{\mu_{en}}{\rho}$ (cm²·g⁻¹)	5.5	0.04	0.025	0.029	0.031	0.032	0.031	0.028	0.022	0.018

Some values for the mass energy-absorption coefficient $\frac{\mu_{en}}{\rho}$, expressed in units of cm²·g⁻¹, in water, or in human tissue, are set out in Table 3.3.

It may be noted that the values for $\frac{\mu_{en}}{\rho}$ remain more or less constant for energies ranging from 200 keV to 1500 keV (around 0.03 cm²·g⁻¹).

In practice, the units used are not SI units, and the above relation then reads:

$$D(x) = 1.6 \cdot 10^{-7} \frac{\mu_{en}}{\rho} E_\gamma \Phi e^{-\mu x}$$

where the absorbed dose D is expressed in mGy, the mass energy-absorption coefficient $\frac{\mu_{en}}{\rho}$ in cm²·g⁻¹, gamma-ray energy E_γ in MeV, the fluence Φ in photons per cm² (i.e. in cm⁻²), the attenuation coefficient μ in cm⁻¹, the material thickness x in cm.

If the fluence rate is used, rather than the fluence, the following relation is obtained:

$$\dot{D}(x) = 5.76 \cdot 10^{-4} \frac{\mu_{en}}{\rho} E_\gamma \dot{\Phi} e^{-\mu x}$$

where the various quantities are expressed in the same units as in the previous expression, except for the absorbed-dose rate $\dot{D}$ expressed in mGy·h⁻¹, the fluence rate $\dot{\Phi}$ in photons·cm⁻²·s⁻¹.

It should be noted that dose calculations, for gamma photons, are carried out for conditions of electronic equilibrium. Which is why the dosimetric standard of interest, in this respect, is the deep dose (i.e. at a depth of 1 cm), for which electronic equilibrium effectively always obtains. In cases where electronic equilibrium is not attained, since the absorbed dose close to the material interface is at a minimum, the above dose calculation would yield an overestimate.

At the same time, at a depth of 1 cm, photon attenuation in the material remains negligible, and the absorbed dose calculation may be consequently simplified:

$$D = 1.6 \cdot 10^{-7} \frac{\mu_{en}}{\rho} E_\gamma \Phi$$

Moreover, at that depth, conditions are free from interactions due to charged particles from the external environment, which as a rule are not highly penetrating, yielding doses at depths shallower than 1 cm (electrons must be endowed with kinetic energies higher than 2 MeV to have ranges of 1 cm or more).

3.1.4.3. Dose–activity relation

A simple relation can be established between the activity A of a point source, expressed in becquerels (Bq), and the dose rate $\dot{D}$ delivered by photons to an individual positioned at a distance d from that source.

3 – Dosimetry

For a point source, the fluence rate $\dot{\Phi}$, expressed in photons·cm^{-2}·s^{-1}, is proportional to the source's activity A, and inversely proportional to the square of the distance d between the source and the individual. The following relation obtains:

$$\dot{\Phi} = \frac{AI_\gamma}{4\pi d^2}$$

The term "AI_γ", in the numerator, corresponds to the number of photons of energy E_γ emitted, per second, by the source, in all directions; while the term "$4\pi d^2$", in the denominator, measures the surface area of a sphere centered at the source, of radius d. This quotient thus measures the number of incident photons from the source, per second, per unit area at a distance d: it thus corresponds to the fluence rate, at that distance. This relation is only valid for distances greater than a few centimeters, and for gamma-emitter point sources.

Substituting this expression for the fluence rate in the expression for the dose rate, we find:

$$\dot{D} = \frac{\mu_{en}}{\rho} E_\gamma \frac{AI_\gamma}{4\pi d^2}$$

Using this relation implies that electronic equilibrium is attained, and that photon attenuation, in the individual considered, is negligible. At the same time, photon attenuation in the atmosphere may be disregarded, especially if the distance d is no more than a few tens of meters.

The general expression, as given above, may be further simplified, if the dose rate, for an individual, is calculated for photons of energies ranging from 200 keV to 1500 keV. In that case, the values for the mass energy-absorption coefficient stand at around 0.03 cm^2·g^{-1} (see Table 3.3), and the relation then reads:

$$\dot{D} = 1.73 \cdot 10^{-5} E_\gamma \frac{AI_\gamma}{4\pi d^2}$$

where the absorbed-dose rate $\dot{D}$ is expressed in mGy·h^{-1}, the gamma-ray energy E_γ in MeV, the source activity A in Bq, the gamma-ray emission intensity I_γ in percent (%) per 100 transformations, the distance d between source and individual in cm.

The relative error levels associated to the use of this relation should be pointed out: the error is less than 10% for gamma-ray energies ranging from 200 keV to 1.5 MeV, 10–30% for gamma-ray energies lying in the 65–200 keV or 1.5–2 MeV ranges. For energies of less than 65 keV, or more than 2 MeV, the error resulting from use of this relation, for the absorbed dose rate, becomes excessively large.

If more precise calculations are undertaken, the values to be used are those established by the International Commission on Radiation Units and Measurements (ICRU), for the fluence-to-dose-equivalent conversion coefficients H^* (10) or H_p (10), depending on photon energy, at a depth of 10 mm in soft tissue (see section 3.3 on operational quantities).

If the source emits a number of photons of different energies, the above formula must be used for each emission energy: the total absorbed dose is the sum of the doses resulting from each emission.

Exercise:

A cesium-137 source with an activity of 100 MBq emits 661-keV gamma rays with a gamma-ray emission intensity of 85%.

Calculate the dose rate in tissue at a distance of 20 cm between source and individual.

Answer:
For 661-keV photons, we can use the dose rate–activity relation:

$$\dot{D} = 1.73 \cdot 10^{-5} E_\gamma \frac{A I_\gamma}{4\pi d^2} = 1.73 \cdot 10^{-5} \times 0.661 (\text{MeV}) \times \frac{100 \cdot 10^6 (\text{Bq}) \times 0.85}{4\pi \times 20^2 (\text{cm}^2)} = 0.2 \text{ mGy} \cdot \text{h}^{-1}$$

3.1.4.4. Inverse-square law

The general expression for the dose rate due to photons from point sources may be further simplified. If we assume a fixed distance, set at 100 cm, the following relation may be used:

$$\dot{D} = 1.37 \cdot 10^{-10} A E_\gamma I_\gamma$$

It should be remembered that this relation takes no account of radiation attenuation in the ambient atmosphere, or in tissue.

This relation makes it possible to calculate the dose rate at a distance of 1 meter. For the purposes of determining the dose rate at any distance from a point source, the inverse-square law, or $\frac{1}{d^2}$ law, comes in as a useful aid. This takes the form:

$$\frac{\dot{D}_1}{\dot{D}_2} = \left(\frac{d_2}{d_1}\right)^2 \quad \text{or:} \quad \dot{D}_2 = \dot{D}_1 \left(\frac{d_1}{d_2}\right)^2$$

where $\dot{D}_1$, $\dot{D}_2$ are the absorbed-dose rates at distances d_1, and d_2 respectively.

Example:

To go back to the previous example:
Remember the experimental data: A cesium-137 source with an activity of 100 MBq emits 661-keV gamma rays with a gamma-ray emission intensity of 85%.

1. *Calculate the dose rate in tissue, at a distance between source and individual of 1 meter.*

   Answer:
   $$\dot{D} = 1.37 \cdot 10^{-10} A E_\gamma I_\gamma = 1.37 \cdot 10^{-10} \times 100 \cdot 10^6 \times 0.661 \times 0.85 = 7.7 \cdot 10^{-3} \text{ mGy} \cdot \text{h}^{-1}$$

2. *Calculate the dose rate in tissue, at a distance between source and individual of 20 cm.*

   Answer:
   $$\dot{D}_2 = \dot{D}_1 \left(\frac{d_1}{d_2}\right)^2 = 7.7 \cdot 10^{-3} \left(\frac{100}{20}\right)^2 = 0.2 \text{ mGy} \cdot \text{h}^{-1}$$

3.2. Protection quantities

The amount of radiation received corresponds to the dose. Dosimetry, or dose estimation, enables the quantification of the physical damage resulting from different types of exposure. The dose cannot be measured in humans, and is estimated for external exposure based on dosimeters described in Chapter 5, "Detection and measurement of ionising radiation".

The notion of absorbed dose is addressed in Section 3.1.1. of this chapter. To reiterate, radiation interacts with the surrounding matter by releasing energy. The amount of energy released is called the absorbed dose. It is expressed in grays (Gy), a unit which corresponds to an energy of 1 joule released in one kilogram of matter (1 Gy = 1 J.kg^{-1}).

3.2.1. Equivalent dose

The probability of stochastic effects occurrence depends not only on the amount of energy absorbed, but also on the nature of the radiation generating the dose. The meaning of the expression "stochastic effects" is explained in Chapter 4, "Biological effects of ionising radiation".

Indeed, the term "radiation" covers both corpuscular radiation (alpha, beta) and electromagnetic radiation (X, gamma). Corpuscular radiation will release its energy entirely during radiation-matter interaction, following a trajectory of variable length that depends on the particle's energy, but also on its size (alpha being much larger than beta), whereas electromagnetic radiation is much more penetrating, and will only be attenuated by interaction in matter. Therefore, for the same absorbed dose, the observed biological effects differ depending on the nature of the radiation.

This difference is taken into account by weighting the absorbed dose by a factor related to the quality of the radiation, called the radiation weighting factor and written "w_R".

The equivalent dose, noted H_T, in a tissue or organ is given by the following expression:

$$H_T = \sum_R D_{T,R} \times w_R$$

where H_T is the equivalent dose for organ T,
w_R is the weighting factor for radiation R,
$D_{T,R}$ is the average absorbed dose in tissue or organ T due to radiation R.

ICRP Publication 60 assigns the weighting factors w_R with each type of radiation, as summarised in Table 3.4.

The unit for equivalent doses is the Sievert (Sv).

The concept of equivalent dose allows the addition of biological effects produced by different types of radiation. It is useful only for regulatory purposes. The values determined for w_R are the result of estimates conducted by comparing the relative biological effectiveness of different radiations in inducing cancer in an organ. Thus, these weighting factors are significant only at low doses of radiation, which lead to stochastic effects. Therefore, the equivalent dose should only be used for exposure to low doses.

Table 3.4. Weighting factor w_R for different types of radiation, ICRP Publication 60.

Radiation	Values of w_R, ICRP Publication 60
Photons of all energies	1
Electrons	1
Neutrons of energy:	
< 10 keV	5
10–100 keV	10
100 keV–2 MeV	20
2 MeV–20 MeV	10
> 20 MeV	5
Alpha particles, fission fragments, heavy nuclei	20

3.2.2. Effective dose

As epidemiological studies have shown, the occurrence of cancers depends on the intrinsic sensitivity of each organ (see Chapter 4, "Biological effects of ionising radiation").

Therefore, each tissue or organ is associated with a weighting factor (w_T) which takes into account the probability of radiation-induced stochastic effects in the organ or tissue. Table 3.5 shows the values of w_T.

This factor allows calculation of the effective dose E.

According to its definition, the effective dose is a hypothetical dose which, administered uniformly to the entire body, would cause the same late-onset damage as all the doses received by the same individual on the different organs separately at different times.

Thus, the concept of effective dose makes it possible to estimate the risk for humans from a measurable quantity, the absorbed dose.

In the case of partial exposure of several organs, the effective dose is defined as follows:

$$E = w_{T1} H_{T1} + w_{T2} H_{T2} + w_{T3} H_{T3} + \cdots$$
$$E = \sum_T w_T H_T$$

The unit for the effective dose is the Sievert (Sv).

Further information

New ICRP recommendations, publication issued in 2007

Since the 1990 recommendations introduced in Publication 60, the ICRP has continued its work. A new publication incorporating new recommendations was issued in 2007. Relatively few changes were made. Some modifications appear in the values assigned to w_R proton = 2 and w_R neutron = continuous function of energy.

Regarding the values of w_T, the risk of radiation-induced cancer was quantified for 8 additional organs and tissues and the w_T were grouped into four families for the sake of simplification ($w_T = 0.12$, $w_T = 0.08$, $w_T = 0.04$ and $w_T = 0.01$). The main change concerns the relative risk for the breast, which was increased from 0.05 to 0.12, and that for gonads, which was decreased from 0.20 to 0.08, as shown in Table 4.4.

Table 3.5. Values of w_T.

Tissue or organ	Weighting factor w_T for tissues CIPR Publication 60
Gonads	0.2
Bone marrow (red)	0.12
Colon	0.12
Lung	0.12
Stomach	0.12
Bladder	0.05
Breast	0.05
Liver	0.05
Oesophagus	0.05
Thyroid	0.05
Skin	0.01
Bone surface	0.01
Other tissues or organs	0.05

3.3. Operational quantities

The body-related protection quantities (effective dose E, equivalent dose in an organ or tissue H_T) have the major drawback of not being amenable to measurement. To meet the requirements of the organizations charged with monitoring workforce exposures, the concept of **operational quantities** was introduced, used to arrive at reasonable assessments of the protection quantities.

These operational quantities involve the following characteristic features:

– they are based on doses at depths of 10 mm, and 0.07 mm, respectively, as measured in the ICRU sphere, or in the human body. The ICRU sphere is a reference sphere, 30 cm in diameter, made of tissue-equivalent material with a density of 1 g·cm^{-3};

– they can be measured at the workplace, by means of external radiation detectors (rate meters, dosimeters), and may be used for individual ambient monitoring purposes at workstations;

- they serve as estimators, yielding as a rule overestimates ("conservative" estimates), for the effective dose, and the organ-related equivalent doses;
- when a variety of radiations, energies, and angles of incidence are involved, the respective quantities relating to each of these are additive.

Three major operational quantities are used:

- two are used for area, or ambient monitoring purposes: the ambient dose equivalent $H^*(d)$, **and the directional dose equivalent $H'(d), \Omega$**;
- the third one is used for individual monitoring purposes: **the personal dose equivalent $H_p(d)$.**

These quantities correspond to the dose equivalent produced at a point located at a depth d in a phantom (e.g. the ICRU sphere), or in the human body, which in turn depends on the energy of the radiation involved, and the geometric conditions pertaining to the exposure (direction of irradiation).

For routine monitoring purposes, the values found for these operational quantities are deemed to provide adequately accurate estimates for the effective dose, and for the dose to the skin, especially if they are lower than the radiological protection limits.

The ambient dose equivalent $H^*(d)$ is the reference quantity for strongly penetrating radiation, for ambient monitoring purposes. $H^*(d)$ provides a good estimator for the effective dose E.

As the recommended depth d, in that case, is 10 mm, this quantity may then be noted $H^*(10)$.

Many detectors used as dose-rate meters are calibrated with reference to $H^*(10)$.

The directional dose equivalent $H'(d, \Omega)$ is the quantity used for low-penetrating radiation, for ambient monitoring purposes. $H'(d, \Omega)$ stands as an estimator for the equivalent dose to the skin H_{skin}. Consequently, the recommended depth, in this case, is 0.07 mm. This quantity may thus be noted $H'(0.07, \Omega)$.

The personal dose equivalent $H_p(d)$ is the quantity used for personal monitoring purposes. Two cases may arise:

- for strongly penetrating radiation, the recommended depth is 10 mm: the quantity is then noted $H_p(10)$, this providing a good estimator for the effective dose;
- for low-penetrating radiation, the recommended depth stands equal to 0.07 mm; the quantity is then noted $H_p(0.07)$, which is a good estimator for the equivalent dose to the skin H_{skin}.

Dosimeters worn on the surface of the body, serving for workforce monitoring purposes, are calibrated with reference to $H_p(10)$ and $H_p(0.07)$; they thus yield good estimates for the effective dose, and the equivalent dose to the skin. Such dosimeters are, as a rule, covered with a tissue-equivalent material.

The depth d, or indeed any thickness of any given material may be stated in terms of the corresponding density thickness, expressed in $g \cdot cm^{-2}$, or in $mg \cdot cm^{-2}$.

Mention should also be made of detectors used for individual monitoring purposes, serving to measure $H_p(3)$, this being an estimator for the equivalent dose to the lens of the eye H_{lens}.

Examples:

- a depth d of 10 mm, in biological tissue, corresponds to a density thickness of 1000 mg·cm^{-2}. One simply needs to multiply the thickness by the material's density:

  10 mm = 1 cm;

  thus: 1 cm × 1 g·cm^{-3} = 1 g·cm^{-2} = 1000 mg·cm^{-2};

- for a depth d of 0.07 mm in biological tissue, the calculation goes as follows:

  0.07 mm = 0.007 cm;

  thus: 0.007 cm × 1 g·cm^{-3} = 0.007 g·cm^{-2} = 7 mg·cm^{-2}.

Figure 3.4 may serve to summarize the relationships between physical quantities, protection quantities, and operational quantities.

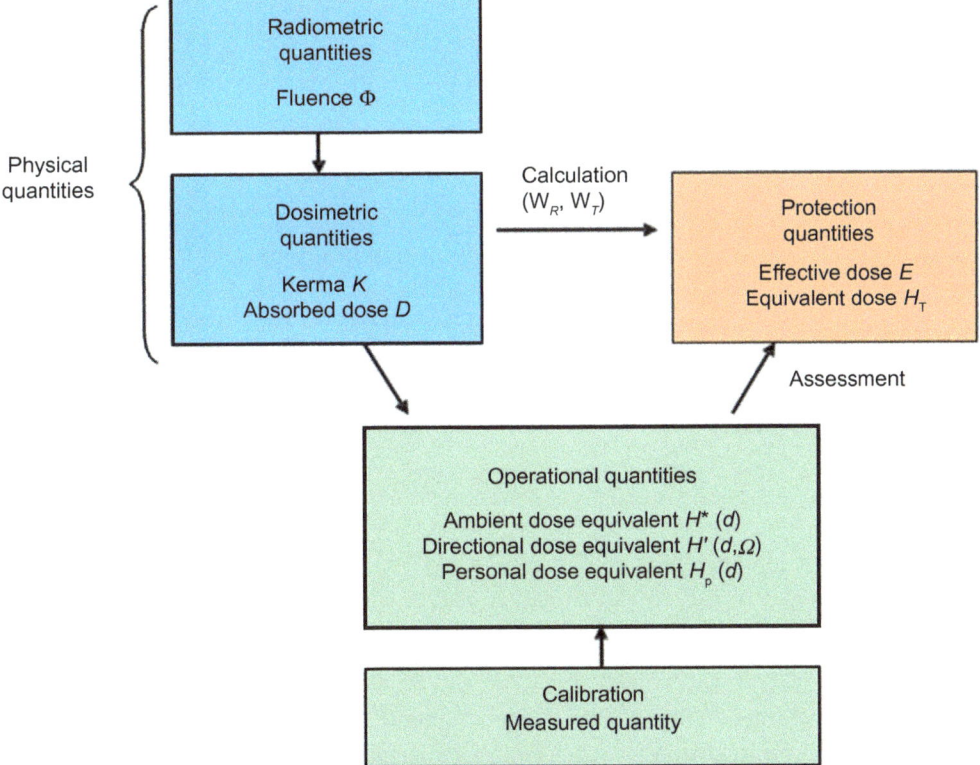

Figure 3.4. Relationships between physical quantities, protection quantities, and operational quantities.

3.4. Check your knowledge

1. **Investigation of the doses generated by a cesium-137 source, and shielding calculations**
Cesium-137 is a β- and γ-emitter, further emitting conversion electrons. The decay scheme for this radionuclide is shown below:

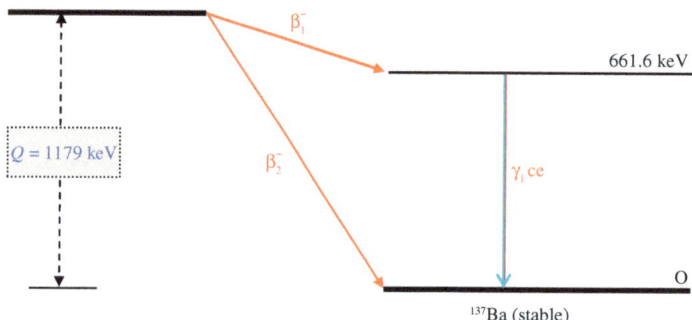

The radiation energies and emission intensities (rounded-off values) are summarized in the table below:

Table 3.6. Low-energy, low-intensity X-rays emitted by the source are disregarded.

Main emissions	E (keV)	%
γ	661	85
β_1 ($E_{\beta\ max}$)	514	94
β_2 ($E_{\beta\ max}$)	1173	6
Electrons (ce K, L, γ) (E_{av})	629	9

An individual is positioned 30 cm away from a 1-GBq cesium-137 point source.

A. **Electron-induced dose**

A1. Calculate the electron fluence rate (i.e. for β_1, β_2, and internal conversion electrons) at a distance of 30 cm from the cesium-137 point source.

A2. Taking into account the average energy of the electrons emitted by the source, which stands at some 180 keV (chiefly due to the energy contribution from the β_1 component), and the mean stopping power at that energy, which is equal to 0.480 keV·μm^{-1}, calculate the mean range for these electrons in tissue.

A3. Calculate the dose rate due to these electrons, at a distance of 30 cm from the cesium-137 point source.

B. **Photon-induced dose**

B1. Calculate the photon fluence rate at a distance of 30 cm from the cesium-137 point source.

B2. Calculate the dose rate due to photons, at a distance of 30 cm from the cesium-137 point source.

B3. Why is the dose rate due to electrons much higher than the dose rate due to photons?

C. Shielding

C1. Bearing in mind that the maximum energy, for electrons emitted by the cesium-137 source, stands at 1173 keV, and that the mean stopping power, at that energy, is equal to 0.218 keV·μm^{-1}, calculate the minimum thickness of plastic material required to stop all of the electrons emitted by the source.

C2. Bearing in mind that the linear attenuation coefficient, for 661-keV photons in plastic, is about 8.6·10^{-2} cm^{-1}, calculate what percentage of photons emerges from the plastic shield. Is this shield sufficient, for the purposes of achieving a sharp reduction in the photon-induced dose rate?

C3. A lead shield is inserted, in order to achieve a steep reduction in the photon-induced dose rate. Bearing in mind that the linear attenuation coefficient, for 661-keV photons in lead, is about 1.17 cm^{-1}, calculate the lead thickness required to bring the dose rate down to 6 μSv·h^{-1}. (The plastic shield may be disregarded, for this calculation.)

C4. Taking this same lead thickness into account, calculate the resulting photon-induced dose rate at a distance of 30 cm from a cobalt-60 point source, having an activity equal to that of the cesium-137 source. To simplify the calculation, an average energy of 1250 keV may be considered, for the photons emitted, together with an emission intensity of 200% (i.e. 2 γ-photons of that energy are emitted for every cobalt-60 decay). At that energy, the linear attenuation coefficient, in lead, is equal to 0.7 cm^{-1}.

Answer:

A. Electron-induced dose

A1. *Calculate the electron fluence rate (i.e. for β_1, β_2, and internal-conversion electrons) at a distance of 30 cm from the cesium-137 point source.*

For a point source located 30 cm away from an individual, the electron fluence rate is equal to:

$$\dot{\Phi}_{electrons} = \frac{A \cdot I_{electrons}}{4\pi d^2} = \frac{1 \cdot 10^9 \times (94\% + 6\% + 9\%)}{4\pi \times 30^2} = 96\,380 \text{ electrons·cm}^{-2}\text{·s}^{-1}$$

A2. *Calculate the mean range for these electrons in tissue.*

The mean range, for these electrons, is equal to:

$$R = \frac{E}{S} = \frac{180}{0.480} = 375 \text{ μm}$$

A3. *Calculate the dose rate due to these electrons, at a distance of 30 cm from the cesium-137 point source.*

We may disregard the radiative stopping power (due to bremsstrahlung), since these are low-energy electrons, and the medium is composed of low-Z elements.

The mean absorbed-dose rate is equal to:

$$\dot{D} = 5.76 \cdot 10^{-4} \frac{\bar{S}_{col}}{\rho} \dot{\Phi}$$

this being the relation set out in the text above, where the mean absorbed-dose rate $\dot{D}$ is expressed in mGy·h^{-1}, the mean collisional stopping power $\bar{S}_{col}$ in MeV·cm^{-1}, density ρ in g·cm^{-3}, the fluence rate $\dot{\Phi}$ in electrons·cm^{-2}·s^{-1}. Therefore:

$$\dot{D} = 5.76 \cdot 10^{-4} \frac{\bar{S}_{col}}{\rho} \dot{\Phi} = 5.76 \cdot 10^{-4} \times \frac{4.8}{1} \times 96\,380 = 267 \text{ mGy·h}^{-1}$$

The absorbed-dose rate, in tissue, thus stands equal to 267 mGy·h^{-1}, yielding an equivalent-dose rate of 267 mSv·h^{-1} (with a radiation weighting factor $w_R = 1$).

The DOSIMEX–B spreadsheet (Calculation of doses induced by ionising radiation – EDP Sciences, 2012) yields the result 246 mSv·h^{-1} (dose-equivalent rate $\dot{H}'(0.07)$) in vacuum. A more realistic calculation, using the same spreadsheet, shows that, at a distance of 30 cm in air, this dose rate comes down to 107 mSv·h^{-1} (100 mSv·h^{-1} in air according to the "Guide pratique radioprotection"). This result is due to the steep degradation undergone by the electron spectrum, particularly with regard to its low-energy component. For that same source, no electron can range further than a distance of 4.7 m in air.

B. Photon-induced dose

B1. *Calculate the photon fluence rate at a distance of 30 cm from the cesium-137 point source.*

For a point source located at a distance of 30 cm from an individual, the photon fluence rate is equal to:

$$\dot{\Phi}_{photons} = \frac{A \cdot I_{photons}}{4\pi d^2} = \frac{1 \cdot 10^9 \times (85\%)}{4\pi \times 30^2} = 75\,160 \text{ photons·cm}^{-2} \cdot \text{s}^{-1}$$

B2. *Calculate the dose rate due to photons, at a distance of 30 cm from the cesium-137 point source.*

In the 200–1500 keV range, the mass energy-absorption coefficient, in tissue, is effectively constant, and the relation thus becomes:

$$\dot{D} = 1.73 \cdot 10^{-5} E_\gamma \frac{A I_\gamma}{4\pi d^2} = 1.73 \cdot 10^{-5} \times 0.661 \times 75\,160 = 0.86 \text{ mGy·h}^{-1}$$

i.e. an equivalent-dose rate of 0.86 mSv·h^{-1}.

The DOSIMEX–G spreadsheet, as the MICROSHIELD® software, both yield the result 1 mSv·h^{-1} (1.1 mSv·h^{-1} according to the "Guide pratique radioprotection").

B3. *Why is the dose rate due to electrons much higher than the dose rate due to photons?*

Electrons, which are directly ionising articles, invariably deposit all their energy across the shallow skin layers (to a depth of 375 μm), whereas photons, as indirectly ionising particles, undergo interaction processes resulting in gradual attenuation over several tens of centimeters, in tissue. For electrons, energy deposition is highly "concentrated", close to the skin surface, resulting in very high dose rates. As regards photons, on the other hand, energy deposition is "distributed" over the entire body, resulting in lower dose rates, over a correspondingly much larger affected volume, however.

It should also be noted, on the other hand, that, while electron-induced dose rates do stand higher, outside a certain skin thickness (375 μm in this case), however, this falls to zero. In contrast, photon-induced dose rates are low, but remain effectively constant across the entire body of an individual.

C. Shielding

C1. *Calculate the minimum thickness of plastic material required to stop all of the electrons emitted by the source.*

The mean range for the electrons, in plastic, is equal to:

$$R = \frac{E}{\bar{S}} = \frac{1173}{0.218} = 5380 \text{ μm} = 5.38 \text{ mm}$$

Interposing a 5.4-mm-thick plastic shield results in all electrons emitted by the cesium-137 source being stopped. The electron-induced dose rate, beyond this shield, is thus equal to precisely zero.

C2. *Calculate what percentage of photons emerges from the plastic shield. Is this shield sufficient, for the purposes of achieving a sharp reduction in the photon-induced dose rate?*

The percentage of emerging photons is equal to:

$$\frac{\dot{\Phi}}{\dot{\Phi}_0} = e^{-\mu x} = e^{-8.6 \cdot 10^{-2} \times 0.54} = 0.95 = 95\%$$

This is not sufficient, for the purposes of achieving a significant reduction in photon fluence. A thicker shield, made of lead, must be interposed.

C3. *Calculate the lead thickness required to bring the dose rate down to 6 $\mu Sv \cdot h^{-1}$.*

The shield thickness may be calculated by taking the natural logarithm of the quantity derived in the preceding relation:

$$x = \frac{\ln\left(\frac{\dot{\Phi}_0}{\dot{\Phi}}\right)}{\mu} = \frac{\ln\left(\frac{860}{6}\right)}{1.29} = 3.9 \text{ cm}$$

A 4-cm-thick lead shield allows the dose rate to be brought down to 6 $\mu Sv \cdot h^{-1}$.

Note: a calculation taking into account the scattering effects taking place within the shield (i.e. by introducing a buildup factor) shows that the actual dose rate thus increases to 12 µSv·h^{-1} (i.e. by a buildup factor of 2). If a dose rate of 6 µSv·h^{-1} is to be effectively achieved, taking the buildup factor into account, a 5-cm lead thickness must be used.

DOSIMEX–G result: for a 1-GBq Cs-137 source at a distance of 30 cm, with a 5-cm lead shield: 6.15 µSv·h^{-1}.

C4. *Calculate the photon-induced dose rate at a distance of 30 cm from a cobalt-60 point source, having an activity equal to that of the cesium-137 source.*

For a point source located at a distance of 30 cm from an individual, the photon fluence rate from cobalt-60 is equal to:

$$\dot{\Phi}_{photons} = \frac{A \cdot I_{photons}}{4\pi d^2} = \frac{1 \cdot 10^9 \times (200\%)}{4\pi \times 30^2} = 176\,840 \text{ photons·cm}^{-2}\cdot s^{-1}$$

If a 4-cm-thick lead shield is used, the photon fluence rate is equal to:

$$\dot{\Phi} = \dot{\Phi}_0 e^{-\mu x} = 176\,840 e^{-0.7 \times 4} = 10\,750 \text{ photons·cm}^{-2}\cdot s^{-1}$$

As is the case for cesium-137, in the 200–1500 keV range, the mass energy-absorption coefficient, in tissue, is effectively constant, and the relation giving the absorbed-dose rate reads:

$$\dot{D} = 1.73 \cdot 10^{-5} E_\gamma \frac{A I_\gamma}{4\pi d^2} = 1.73 \cdot 10^{-5} \times 1.250 \times 10\,750 = 0.23 \text{ mGy·h}^{-1}$$

The absorbed-dose rate, in tissue, thus stands equal to 0.23 mGy·h^{-1}, i.e. an equivalent-dose rate of 0.23 mSv·h^{-1} (with a radiation weighting factor $w_R = 1$). Compared to what was found for the cesium-137 source, the resulting dose rate is far higher (by a factor 40). This discrepancy is due, on the one hand, to the cobalt-60 photon-fluence rate, which is higher (by a factor 2), together with a higher photon energy (again by a factor 2), and, most crucially, the lower attenuation coefficient for these higher-energy photons, in lead (10 times less attenuation). It should be noted that a shield which proves adequate for one source is not necessarily suitable as regards affording protection against another source, of a different type, albeit of equal activity. To sum up, care should be taken to cater for any change in the source being used, and not to assume that if a set of good conditions is adequate for one source, the same will prove sufficient with another source.

DOSIMEX–G results:

- *Naked source: 3400 µSv·h^{-1};*
- *With a 4-cm lead shield: 293 µSv·h^{-1} (no buildup), 558 µSv·h^{-1} with buildup;*
- *To achieve, here again, a dose rate of 6 µSv·h^{-1}, a 10-cm lead thickness is now required – with no account taken of buildup – 12 cm if buildup is taken into account.*

2. The absorbed dose is:

a. the number of times a subject was exposed to radiation

b. the amount of radiation received by the worker during his/her activity

c. the amount of energy received by an individual per unit mass of traversed matter

d. all of the above

Answer: c.

3. The SI unit for the absorbed dose is:

a. C/kg

b. gray

c. sievert

d. rem

Answer: b.

4. In radiation protection, the quantity that takes into account the nature of the radiation is:

a. exposure

b. the absorbed dose

c. the equivalent dose

d. the effective dose

Answer: c.

5. In radiation protection, the quantity that takes into account the characteristic radiosensitivity of each organ is:

a. exposure

b. the absorbed dose

c. the equivalent dose

d. the effective dose

Answer: d.

6. The SI unit for equivalent dose and effective dose is:

a. C/kg

b. gray

c. sievert

d. rem

Answer: c.

7. In Figure 3.5, two characters are facing each other. One holds a large number of balls which he throws towards the other. Some of these balls reach the second character.

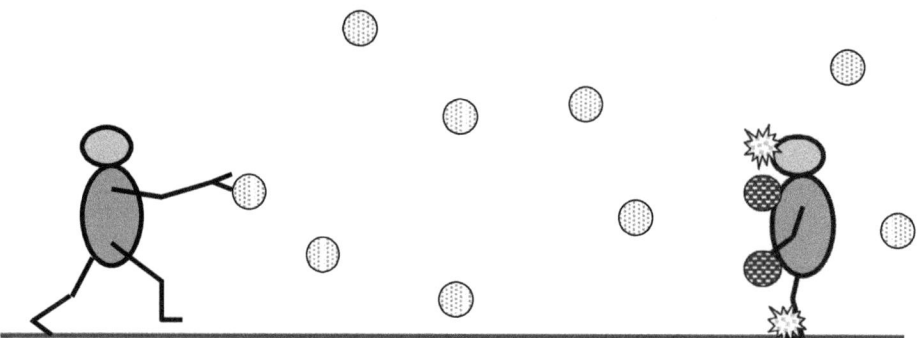

Figure 3.5. Image of the quantities and units used in radiation protection.

With which quantities used in radioactivity or radiation protection would you associate the following events: number of balls thrown, number of balls hitting the second character, location of the ball impacts and pain associated with the ball impacts?

Answer:

- The number of balls thrown represents a number of disintegrations per second; therefore, it is an image of the activity of a source, measured in becquerel (Bq).
- The number of balls hitting the second character represents the absorbed dose in grays (Gy).
- The impacts of the balls on the body represent the equivalent dose and the effective dose in sieverts (Sv).

4 Biological effects of ionising radiation

Christine Jimonet, Henri Métivier

Introduction

In 1895 Wilhelm Röntgen observed that accelerated electrons striking a metal plate in a vacuum glass bulb caused a nearby fluorescent screen to illuminate, even when the bulb was wrapped in black paper.

He concluded that the bulb emitted an unknown radiation capable of passing through glass, paper and air, which he called X-rays. Interposing his hand between the bulb and a photographic plate, he saw the image of the bones in his hand appear on the plate: X-rays are also capable of passing through the human body. The photography toured Europe and physicians immediately understood the value of such a physical phenomenon: it was the birth of radiology.

However, in 1900 Henri Becquerel observed a red spot on his skin beneath his waist-coat pocket, where he carried a tube containing radium. In 1902 the first cancer associated with the use of X-rays was described in an employee who, for several years, had checked on his own hand that the tubes operated properly. Many other cancers occurred among radiologists, physicians and physicists who used X-rays intensively, and the need to regulate their use was thus recognised. In the early 1920s, radiation protection regulations were prepared in several countries, but it was not until 1925 that the first International Congress of Radiology took place and considered establishing international protection standards. Three years later, at the International Congress of Radiology in Stockholm, two international non-governmental commissions were established: the International Commission on Radiation Units and Measurements (ICRU), and the International Commission on Radiological Protection (ICRP), being responsible for defining the rules of radiation protection. These two commissions continued their work and current regulations about radiological protection relies heavily on their recommendations.

For a radiation source which is located outside the human body, it is necessary that the radiation is able to reach the target molecules. In other words, the radiation must be sufficiently penetrating. This is not the case for radioactive substances that have penetrated into the body through inhalation, ingestion, injury or skin transfer, as radionuclides are then in direct contact with the cells and their constituents.

As explained in Chapter 2, "Interaction of ionising radiation with matter", ionising radiation loses energy by ionising and exciting the atoms as they pass through cells. Biological effects that are observed at the body level are thus a result from molecular lesions induced by ionisation, which leads to disruption of chemical bonds of biological molecules.

Two approaches are used to study the biological effects induced by the degradation of these biological molecules: epidemiology and experimentation. Epidemiology is the observation of the effects on populations that have been exposed naturally or artificially to radiation. Examples of such populations include the survivors of the Hiroshima and Nagasaki bombings, people exposed to radiation during the nuclear accident in Chernobyl, uranium mine workers... Through experimentation researchers can, among other things, observe the damage and disruption caused to the DNA molecule, the carrier of genetic information, by ionising radiation or analyse the repair mechanisms that become activated in cells with damaged DNA.

4.1. Molecular effects of interaction with ionising radiation

As is any material, the human body is composed of atoms, predominantly hydrogen, oxygen, carbon and nitrogen, which are grouped into different kind of molecules. Water molecules are the main constituent of all cells of the human body. Besides water, organisms are also composed of other large families of molecules including lipids, proteins, carbohydrates, nucleic acids. These molecules are organised to form the cells, which are small compartments bounded by a membrane that contain an aqueous solution of chemical compounds. The body is composed of billions of cells with different roles, and cells of similar structure and function are grouped together to form tissue: nervous tissue, muscle tissue...

Very often, several different types of tissue make up an organ associated with a specific function. A collection of organs performing complementary, interacting functions define a system: circulatory system, digestive system... Ultimately, all systems work together within the organism.

All biological consequences of ionising radiation on living tissues are a result of the interaction with atoms in a process called 'ionisation'. Ionising radiation has sufficient energy to remove one orbital electron from an atom, thereby creating an ion pair.

There are two mechanisms by which radiation ultimately affects cells: either by direct ionisation of the target molecule, or indirectly by the production of free radicals which may ultimately affect the target molecule.

During direct ionisation, radiation transfers energy directly to the atoms of critical cellular components such as deoxyribonucleic acid (DNA). The mean number of inactivated or damaged molecules is directly proportional to the radiation dose. If a cell is exposed to ionising radiation, the probability of directly interacting with DNA is low, as it makes up only a small part (1%) of the cell. It is widely accepted that DNA is the critical target molecule of ionising radiation, even if the modification of membrane lipids and some amino-acids comprising proteins can disrupt cell life as well.

The main constituent of all cells of the human body is water (up to 80%). Therefore, most of the energy produced by ionising radiation leads to water radiolysis, which results in free radical production ($OH°$, H_2O_2) that ultimately can lead to indirect DNA damage. It should be noted that the formation of free radicals, highly reactive chemical species, is not specific to the action of ionising radiation. The action of UV light or metabolism can

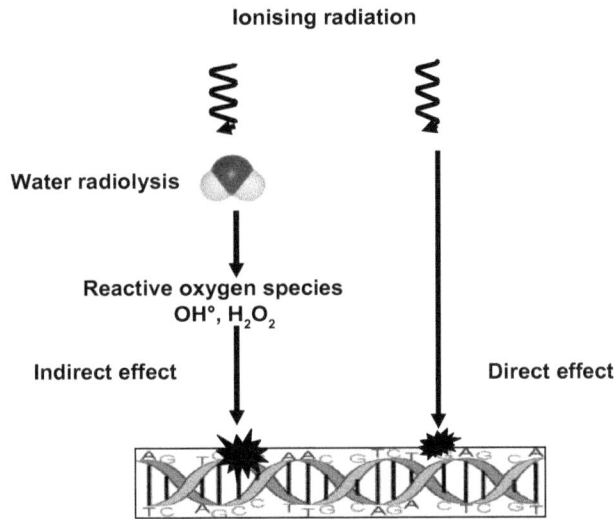

Figure 4.1. Direct and indirect effect of ionising radiation on DNA.

also lead to the formation of these radicals, which are partly responsible for the phenomena related to aging.

The molecular interaction of radiation by direct and indirect effect is illustrated in Figure 4.1.

Further information

The cell and DNA

Based on *Fédération Nationale des Centres de Lutte Contre le Cancer (National Federation of French Cancer Centres)*
Bernard Hœrni.

The word 'cell' comes from the Latin cellula, meaning "a small room" and is the smallest unit of all living organisms including animals or plants.

Among its components, as shown in Figure 4.2, we can distinguish a nucleus, a cytoplasm and a membrane that separates the interior of all cells from the outside environment. The nucleus contains most of the cell's genetic material, organized as multiple long linear DNA, represented by a set of genes (20 000-25 000) in a form more condensed than that in a computer chip (see Figure 4.3). DNA controls the production (synthesis) of proteins and all substances required for the life of the cell. The cytoplasm contains a genuine biological factory with numerous "workshops" dedicated to perform

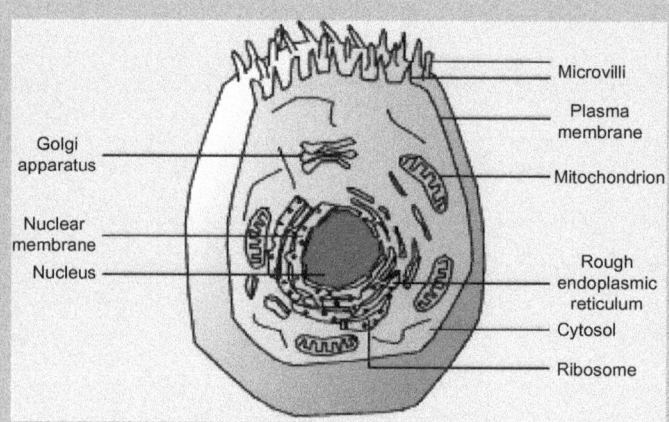

Figure 4.2. Schematic representation of a cell (drawing: Marion Solvit).

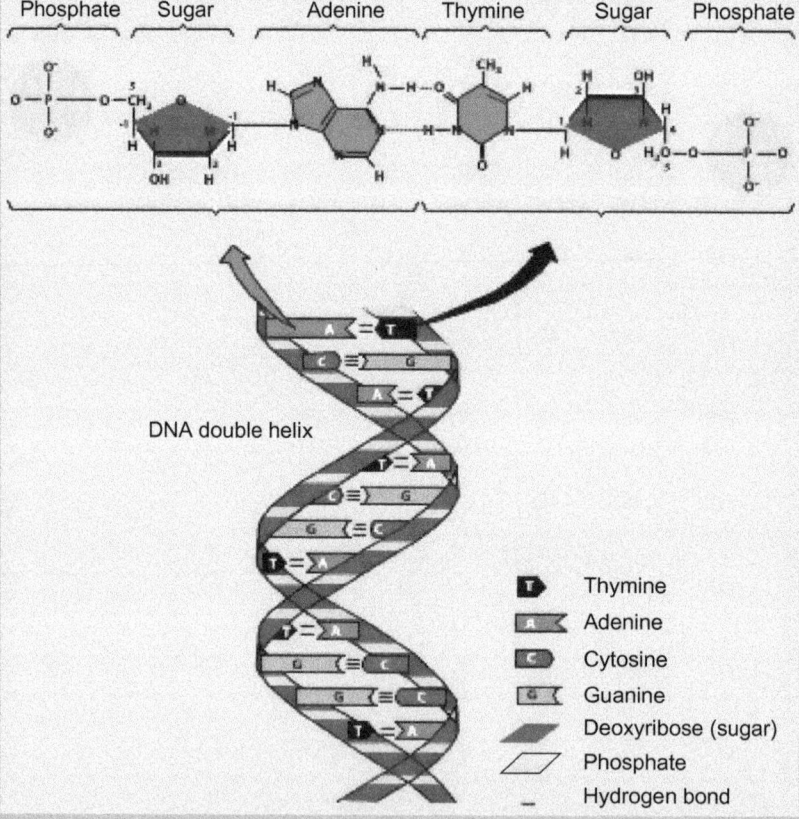

Figure 4.3. Representation of the DNA double helix (drawing: Marion Solvit).

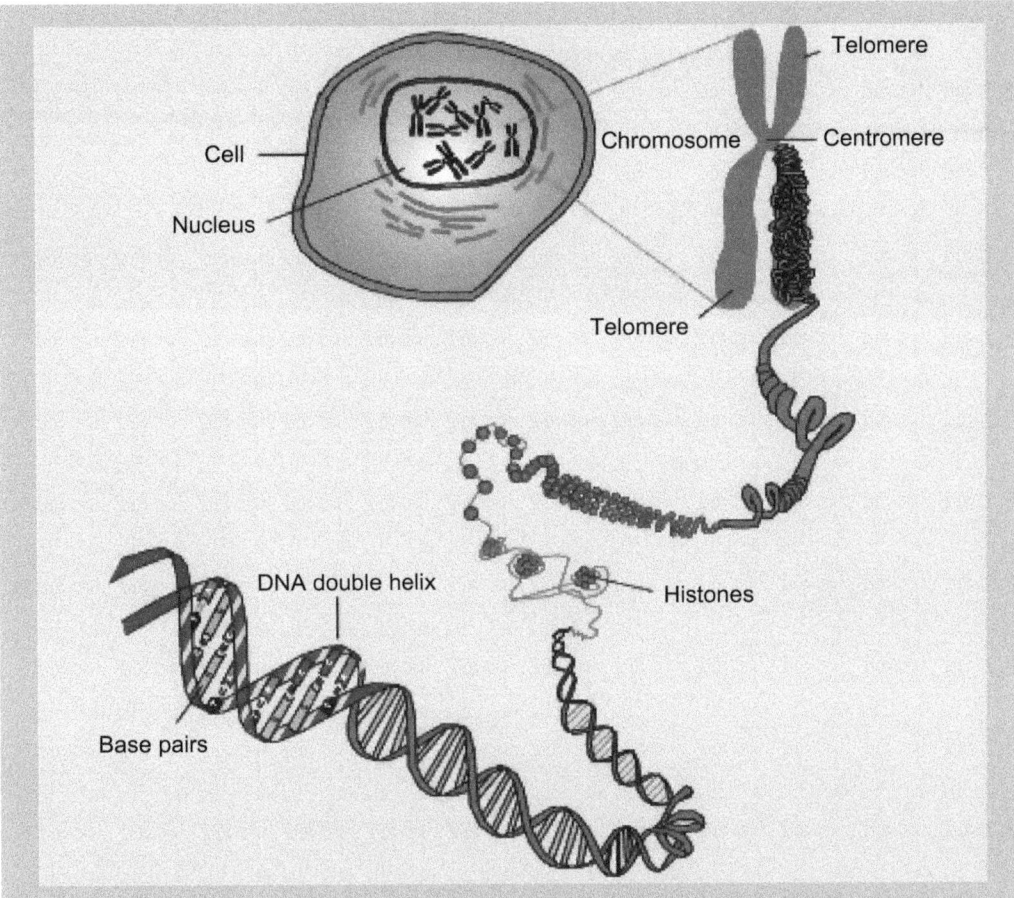

Figure 4.4. DNA condensation during the formation of the chromosomes (drawing: Marion Solvit).

different tasks. Some, according to the information contained in the DNA, synthesise proteins from their elementary components, i.e. amino-acids. Some of these proteins act immediately within the cytoplasm as enzymes that facilitate (catalyse) biological reactions. Others are excreted from the cell to act on other cells, adjacent or remote, as hormones do. Yet others are responsible for cellular function, such as the contraction of muscle fibres. Rather than being a barrier, the cytoplasmic membrane is the site of exchange. It releases products synthesised within the cell in order to act elsewhere, or to be eliminated. The membranes of certain cells exhibit structures that allow them to attach themselves to their neighbours. This provides coherence, and sometimes extreme strength to a tissue: this is the case for the squamous epithelium of the skin, which is flexible and yet among the most resistant types of tissue of the organism.

4.2. Cellular effects, consequences of molecular effects

Depending on the complexity of the DNA modifications caused by direct or indirect effect, different cellular effects may occur (Figure 4.5).

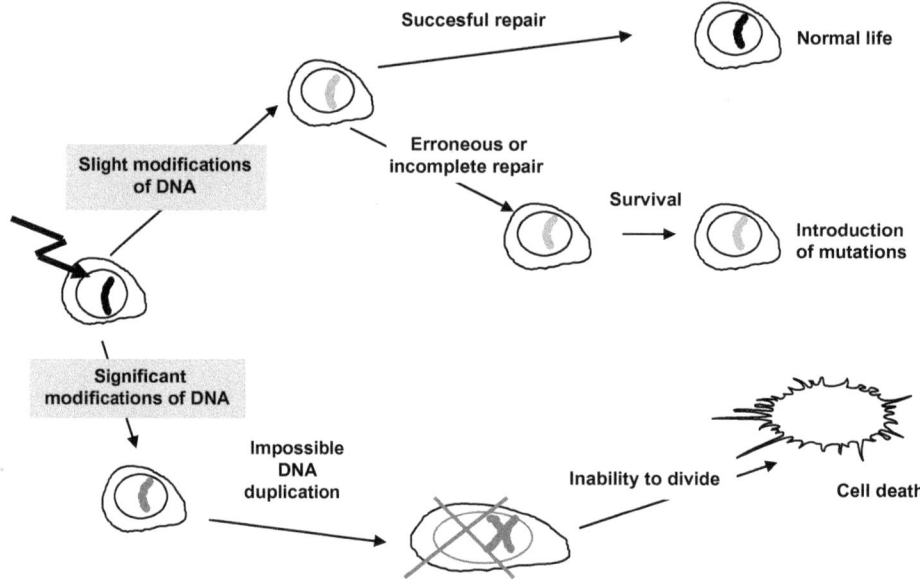

Figure 4.5. Cell fate following DNA modifications.

First, let us consider the case of a cell whose DNA is slightly damaged. It is very likely to recover a normal life, as alterations in the DNA molecule are inherent in the life of the cell. Living organisms have evolved in a world constantly exposed to radiation. In addition, cellular DNA is exposed daily to potential damage from numerous physical or chemical agents. The cell therefore possesses a set of tools to repair these molecular alterations. If the damage is fully repaired by the enzymatic DNA repair machinery that is present in the cell, cells will normally survive. However, if the DNA repair is incomplete or incorrect, the genetic information of the cell is altered, thus creating a mutated cell.

Finally, cell death (immediate or during the next cell division) can be the ultimate fate of a cell whose nucleus has severe DNA damage. Cell death may occur through two different mechanisms: necrosis and apoptosis. Necrosis leads to the breakdown of the cellular structures and the release of its more or less harmful contents into the surrounding environment, thereby causing an inflammatory reaction.

Apoptosis is the process of programmed cell death by self-destruction, the cell fragmenting itself into smaller segments which are not harmful to the environment. This programmed cell death is in fact a protective mechanism that prevents the organism from retaining cells whose genetic code has been modified.

So far, we only focussed on the effects of irradiated cells. However, in certain circumstances, ionising radiation can affect cells other than those directly irradiated. These effects are called "non-targeted effects" and explain changes in non-irradiated cells or in the offspring of irradiated cells. The 'bystander effect' is the phenomenon in which non-irradiated cells exhibit irradiated effects as a result of signals (chemical factors) received from neighbouring irradiated cells. Genomic instability is the phenomenon in which DNA damage becomes visible in the daughter cells of irradiated cells. It refers to an increased tendency of alterations in the genome during the life cycle of a cell.

Further information

Mutations and repair systems

Based on "Génétique moléculaire : principes et application aux populations animales" (Molecular Genetics: principles and application to animal populations), INRA, 2000.

The basal mutation rate of DNA reflects the inevitable errors during DNA replication or repair. Indeed, the fidelity of DNA polymerases is not absolute, and replication errors occur with a frequency of about 10^{-9} per incorporated nucleotide, i.e. about one mutation per cell division throughout the genome. In a human lifetime, the organism performs a number of cell divisions on the order of 10^{17}; the genome thus undergoes about 10^{17} somatic mutations. In the germ line, the number of divisions required to produce respectively an egg and a spermatozoon is estimated as 24 and (30+23n+5 where n = age in years-15). Each egg thus carries about 24 localised mutations and each spermatozoon of a 25 year old man comprises almost 265 mutations associated with replication errors.

Furthermore, DNA is subject to spontaneous chemical modifications, mainly hydrolysis reactions:

- spontaneous depurinations due to the unstable N-glycosidic bond connecting the puric bases to the poly-deoxyribose-phosphate skeleton. This bond may be hydrolysed, and the purine (A or G) replaced with a hydroxyl group on the 1' carbon. These are very common events: 5000 puric bases are lost every day by each human cell;
- tautomeric transitions (-NH2 → =NH or C=0 → =C-OH), in particular deaminations with adenine to hypoxanthine transitions, guanine to xanthine transitions, and cytosine to uracil transitions. The latter case occurs with a frequency of approximately 100 bases per day per cell in humans;
- transversions, which entail the substitution of a pyrimidine (C or T) by a purine (A or G), or vice versa.

Genomic DNA can also be altered by exposure to multiple mutagens, endogenous (free radicals) and exogenous (UV radiation), which can induce the formation of covalent bonds between two adjacent pyrimidines (CC, CT or TT) on the same DNA strand. Most of these modifications (breaking of a strand, creating covalent bonds between complementary bases, chemical modification of a base, base excision...) are generally caught and corrected by the cellular enzyme repair systems, except in the case of deamination of a methylated cytosine. Indeed, in this case, the product formed (5-methyl uracil) is the equivalent of thymine, which is not perceived as foreign to the DNA and therefore not repaired. The accidental transition C → T is thus usually fixed. Regions rich in methylated cytosines, in particular CG doublets, are therefore vulnerable and constitute mutational hot spots.

If the repair is imperfect, the accidental mutation can be fixed. Mutations that occur in the germ line can be transmitted to subsequent generations, potentially generating a polymorphism, whereas somatic mutations affect only the individual in which they occur. In somatic cells, if the mutation has no effect on the operation of the cell, the mutant cell clone is diluted in the population and the mutation remains unnoticed. If the mutation is deleterious, the clone spontaneously expires. If instead, the mutation provides a selective advantage, the mutated clone can proliferate, as in the case of cancer.

4.3. Non-stochastic or deterministic effects

Observations grouped under the term "non-stochastic effects" relate to the types of cells making up tissue.

Healthy tissue is composed either of a single differentiated cell type, or of several compartments of different cell maturation. In the latter case it is standard to distinguish the stem cells compartment, made up of young cells with high mitotic activity, the maturing cells compartment and the functional compartment of adult cells without mitoses.

As shown in Table 4.1, the lifespan and the amount of cells eliminated daily are two characteristics that depend on the tissue to which the cells belong.

Table 4.1. Cell renewal depending on the type of cells.

Cells	Lifespan	Losses
Muscle	Same as the individual	A few hundred/day
Blood	Erythrocytes: 120 days	200 billion/day
Intestinal mucosa	about 3 days	100 billion/day
Epithelial	about 30 days	1.5 g/day

To ensure normal function, the organism must continually produce blood cells, epithelial cells and intestinal cells, and it does so by relying on the division of stem cells from the bone marrow, the skin and the intestinal crypts compartments, respectively.

Because they divide so intensively, stem cells are the most radiosensitive cells of the human body. As early as 1906, Bergonié and Tribondeau stated that the less differentiated its composite cells are, the greater their proliferation potential, and the faster their division, the more radiosensitive a tissue is.

The functioning of most organs and tissues is not affected by the loss of a moderate number of cells. However, if the number of cells that are destroyed is too high, or if vital cells such as stem cells in bone marrow, skin or intestinal crypts are destroyed, a loss of function will be observed in the tissue.

Deterministic effects are associated with the concept of irradiation threshold dose: below this dose, no effect occurs; once the threshold has been exceeded, the effect will always occur. Correlative to this threshold dose is the notion of severity of the effect, which increases with dose received.

In other words, because deterministic effects result from **the death of a large number of cells**, the destruction of a few cells does not affect the physiological characteristics of an organ, whereas the destruction of a large number of cells leads to organ malfunction, hence the concept of **threshold dose**.

4.3.1. Effects of localised irradiation

4.3.1.1. Skin

Illustrated in Figure 4.6, the epidermis is a dynamic system consisting of a basal layer of stem cells, a few layers of epithelial cells, and a layer of dead desquamating cells.

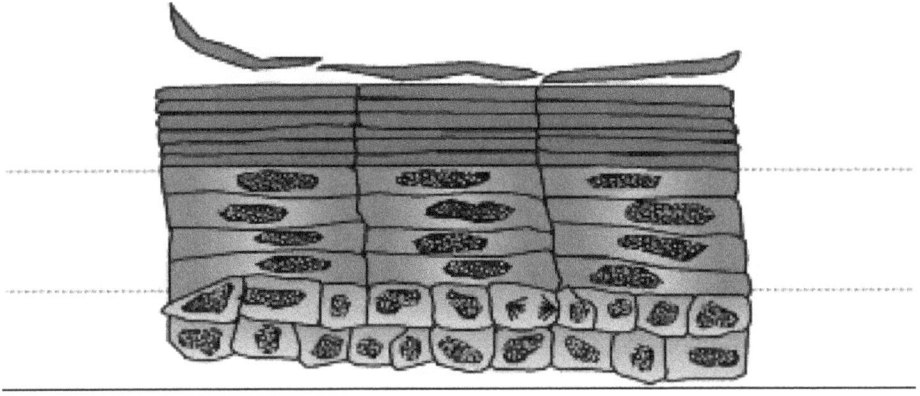

Figure 4.6. Section of the epidermis (drawing: Marion Solvit).

Radiation-induced lesions of the skin are called radiodermatitis.

A dose of 1 Gy delivered locally induces the death of epidermis basal cells, yet without observable damage apart from an atrophy of the skin.

A dose of 5 Gy leads to the formation of an erythema (sunburn, redness of the skin) and the desquamation of the epidermis' upper layer.

Above 10 Gy, an exudative epidermitis occurs with ulceration due to damage to the basal cells and to the vascular system.

4.3.1.2. Eyes

The crystalline lens is the most sensitive part of the eye, and exposure to ionising radiation can lead to radiation-induced cataracts, resulting in lens opacification.

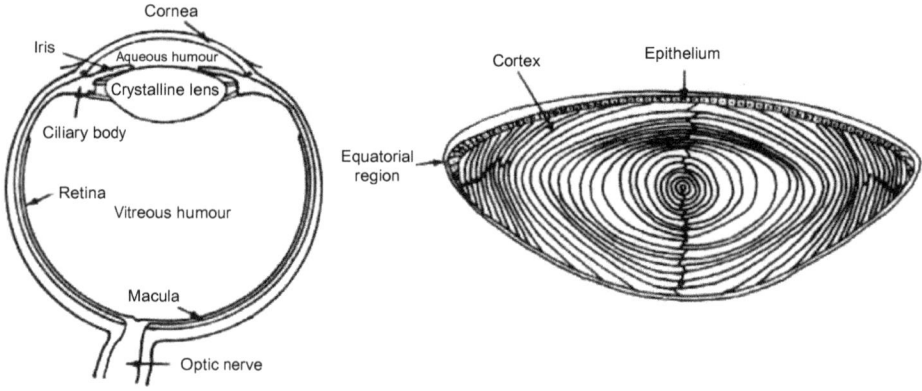

Figure 4.7. Section of the eye and crystalline lens.

Figure 4.7 shows the structure of the eye and crystalline lens. The lens is biconvex and located behind the iris. Its purpose is to focus light on the retina. To this end, it is composed of very specific cells, fibrous, transparent and highly elongated, which conduct light. These cells are anucleate and contain no organelles; they are composed only of their plasma membrane and cytoplasm, and are known as lens fibres. These fibres are created, after differentiation, by the division of epithelial cells which belong to the stem cell layer located on the exterior side of the crystalline lens.

The most radiosensitive area in the crystalline lens therefore corresponds to the epithelial layer where cell division occurs continuously, producing the future lens fibres.

If, as a result of exposure to ionising radiation, too many epithelial cells are damaged, the structural organisation of the lens fibres will be altered, resulting in opacities.

Until recently, it was agreed that in the case of a single exposure, a detectable but non evolutive opacity would occur for an absorbed dose of 1 to 2 Gy, the threshold dose leading to a cataract being of 2 to 5 Gy. However, recent studies have reported a greater radiosensitivity of the lens than initially estimated, leading the ICRP to issue new recommendations. The ICRP now considers 0.5 Gy, a significantly lower value, to be the absorbed dose threshold leading to the occurrence of a cataract. In addition, the Commission now recommends that the equivalent dose limits for the crystalline lens be lowered to 20 mSv per year, with the possibility of averaging to 100 mSv over 5 years without exceeding 50 mSv/year.

The estimated latency between exposure to ionising radiation and cataract occurrence does not seem to have been modified, and would therefore be 10 years.

4.3.1.3. Reproductive organs

Table 4.2 below shows the threshold doses in cases of single, localised exposure of the reproductive organs leading to temporary or permanent sterility.

The large difference in radiosensitivity between the two types of germ cells is explained by the genesis of each cell types.

Table 4.2. Threshold doses for reproductive organs.

Reproductive organs	Man	Woman
	Threshold dose (Gy)	Threshold dose (Gy)
Temporary sterility	0.3	2
Permanent sterility	5	7

Indeed, because spermatozoons have an average lifespan of 72 hours, they are continually produced by intense division and differentiation of "precursor" cells, while for women, the total stock of oocytes is set at birth.

4.3.1.4. The embryo

The radiosensitivity of the embryo and human foetus varies according to its developmental stage.

If exposure occurs before implantation of the egg, i.e. before the 9th day after conception, effects follow an "all-or-nothing" pattern: irradiation either leads to the loss of the embryo, or implantation occurs and the exposure will have no impact on the development of the embryo.

At the organogenesis stage (from the 9th day to the 60th day), tissue radiosensitivity is at its highest and exposure may cause malformations.

During the foetal stage, the nervous system remains radiosensitive; irradiation can cause mental retardation, especially between the 60th and the 110th day.

In Hiroshima and Nagasaki, no malformations or mal-development of the embryo were observed in patients who had been irradiated in utero at doses below 0.5 Gy. The value of 0.1 Gy is generally accepted as quasi dose threshold (*ICRP Publication 84, Pregnancy and Medical Radiation, 2001*).

4.3.2. Effects of a single, global and homogeneous irradiation of the entire organism

Descriptions of pathologies are based on observations following accidental irradiation, or whole body irradiation for medical purposes (organ transplants, leukaemia).

The clinical picture results from the destruction of the most radiosensitive cell types (bone marrow, intestinal mucosa, skin, testes).

Threshold dose values may vary depending on the author, and are given as an order of magnitude.

Above 0.3 Gy, a temporary change in the blood count is observed.

Above 1 Gy, the blood stem cells residing in the bone marrow, called hematopoietic cells, are affected, known as hematopoietic syndrome.

Of all blood cells except red blood cells, lymphocytes are the most radiosensitive cells. This type of blood cells is responsible for the immune defence system; they are followed by white blood cells (polynuclear leukocytes), which play a role in the destruction of infectious germs, and finally by platelets (responsible for coagulation). This syndrome leads to decreased resistance to infections and haemorrhage.

Higher doses (>6 Gy), will also result in gastrointestinal syndrome leading to the destruction of the small intestine's cellular lining, which is made of villi. Absorption phenomena are totally altered and complicated by the risk of infection.

Very severe whole-body exposure (20-40 Gy) is characterized by a deteriorating state of consciousness with eventual coma and death from respiratory failure within 1 to 2 days (neurovascular syndrome).

50% of the patients die in the absence of any treatment for an average value of whole body exposure of 4.5 Gy: this value therefore corresponds to the median lethal dose.

This information is summarised in Figure 4.8.

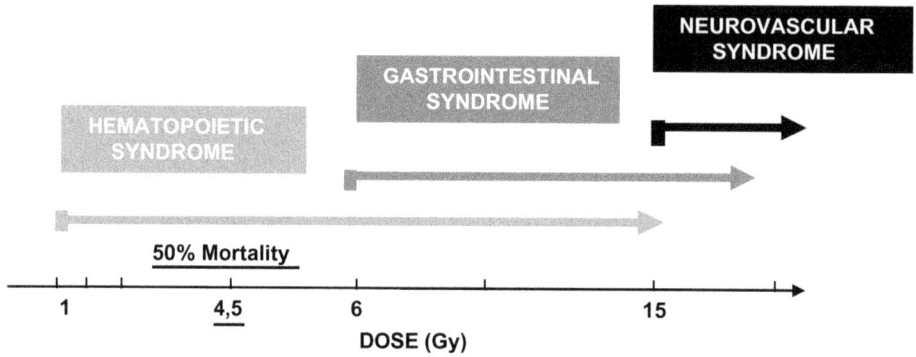

Figure 4.8. The three syndromes in case of whole-body radiation exposure.

In case of acute accidental irradiation, the clinical picture corresponds to an irradiation syndrome in which four phases can be identified:

- the immediate or prodromal phase, characterised by nausea and vomiting: the time of onset and severity of these symptoms vary with the dose;

- the latent phase, during which symptoms disappear: the higher the dose, the shorter this period is;

- the critical phase, during which the characteristic signs of each of the syndromes described above develop;

- the last phase, which can be either fatal or a recovery phase.

In case of an incident/accident, it is important to assess the received dose, as the severity of the pathology depends on this dose. This evaluation can be performed using:

- legal and physical dosimetry, although it is often difficult to use in an accident situation;

- clinical examination: indeed, for each symptom, similar to lethal doses, there are average doses related to certain effects in exposed persons. These assessments should take into account individual variations;

- biological dosimetry including haematological, biological and cytogenetic assessments, the latter analysing the number and shape of the chromosomes;

- dosimetric re-enactment by numerical simulation or by experimental reconstitution using dummies containing dosimeters.

4.3.3. Characteristics of deterministic effects

Effects are called non-stochastic or deterministic (i.e. inevitable) when they have the following characteristics:

- they are recognisable and distinguishable from other pathologies;

- there is an exposure threshold, dose value above which lesions or pathologies will be observed in all subjects, hence the term deterministic;

- the severity of the damage increases with the dose;

- they appear rapidly (days, months).

When regulations are observed, the normal practice of an activity utilising ionising radiation is not likely to cause deterministic effects. Such effects are only observed following accidental exposure (accident in a facility, inappropriate manipulation of a highly active radioactive source) or intentional exposure of patients for therapy (radiotherapy).

Further information

Radiation accidents over the last 60 years
Jean-Claude Nenot
J. Radiol. Prot. 29 (2009) 301–320

Radiotherapy

Radiotherapy is a targeted treatment modality for cancers, which uses the energy deposited by radiation to kill cancerous cells thereby inhibiting their ability to proliferate while sparing the healthy surrounding tissues.

The dose delivered in radiotherapy is measured in Grays (Gy). The radiation therapist prescribes a dose to be delivered in a given area (usually the tumour) and the fractionation to be used (dose per day). If necessary, he/she defines the appropriate dose constraints not to exceed in the surrounding areas (organs at risk). The prescribed dose and fractionation depend on the location and nature of the disease. Generally, a dose of 45-80 Gy is delivered to the target in fractions of 2 Gy/day (order of magnitude).

4.4. Stochastic effects

Stochastic effects are the result of unrepaired or poorly repaired damage to the DNA molecules, leading to mutated cells that did not undergo apoptosis, and were not eliminated by the immune system.

If the mutation affects the genome of somatic cells (cells of the body), a cancer may develop.

If the mutation occurs in the genes of germline cells (oocyte, spermatozoon), the effects will be genetic and appear only in the offspring of the exposed individual if the offspring was conceived from the mutated parental sex cell. This is known as genetic or hereditary effects.

It should be noted that the monitoring of exposed populations (Hiroshima-Nagasaki) has not revealed an excess of hereditary effects, and this remains the case with the second generation, currently being monitored. The teratogenic abnormalities observed in these populations are abnormalities in brain development: microcephaly with or without intellectual deficit.

Care should be taken not to confuse genetic and teratogenic effects. Teratogenic effects can be either deterministic or stochastic, such as for an already born child, and they are related to the child's irradiation through the mother. They are not particularly transmissible.

Cancer and genetic effects, on the other hand, constitute what are called stochastic effects, also known as random effects since they are related to the field of probabilities.

Characteristics of stochastic effects

Effects are called stochastic when:

- they are not specific to the action of radiation;

- they appear only in certain individuals;

- the probability of their occurrence increases with the dose;

- their severity is independent of the dose;

- they have a late onset, about 5 to 10 years for leukaemia, 20 to 50 years for most other cancers and at least a generation for genetic effects;

Table 4.3. Characteristics of the two types of biological effects.

Deterministic effects (inevitable)	**Stochastic effects (random)**
	Cancers and hereditary genetic effects.
– Only at **high doses**	– Even with **low doses**
– Existence of a **threshold**	– **No threshold**
– **Early onset**	– **Late onset**
– **Severity** depends on the dose	– **Frequency** depends on the dose

- as a precaution, the ICRP and regulators consider that the relationship between the frequency of occurrence of these effects and the dose is linear, without a threshold.

4.5. Summary

The genesis of the events related to biological effects is summarised in Figure 4.9, from the initial interaction to the possible transition into pathology. Note the time scale associated with the different events.

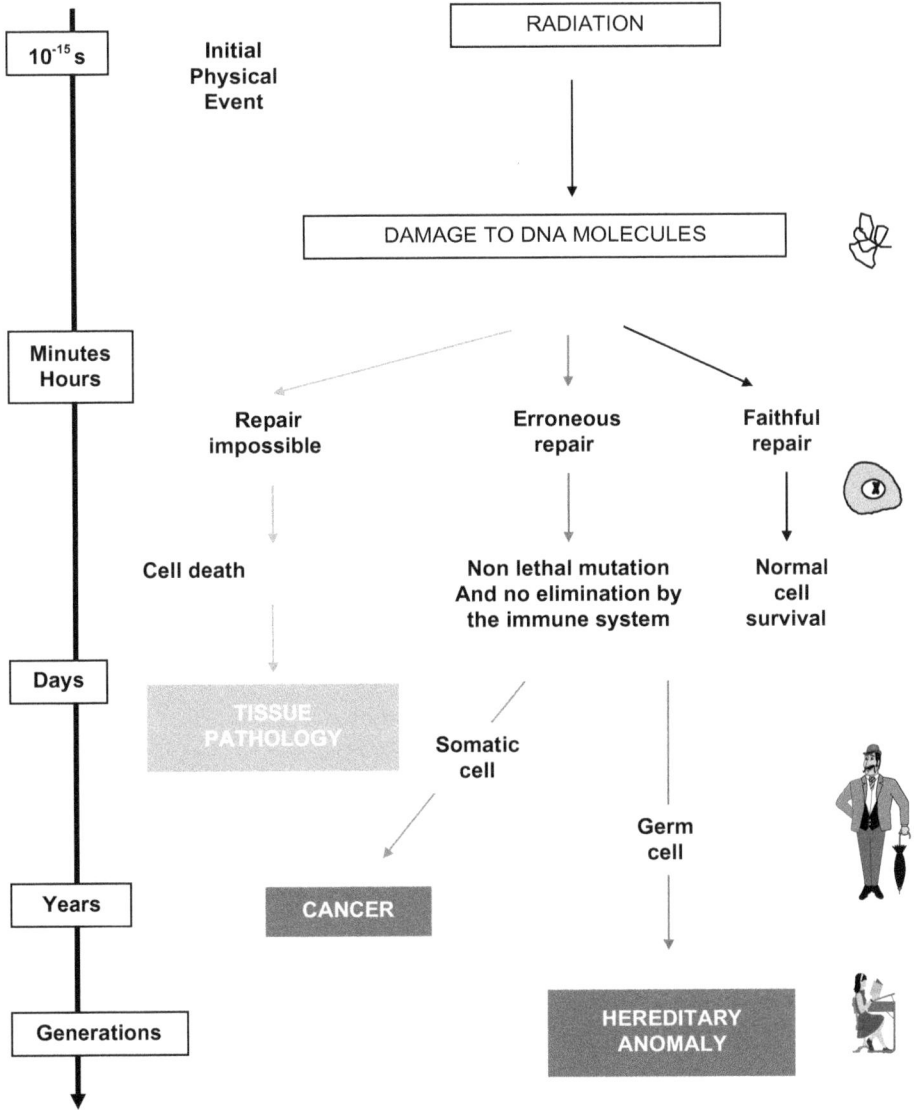

Figure 4.9. Overview of the genesis of biological effects.

4.6. Risk assessment

4.6.1. Carcinogenic effects

Excluding accidental situations, entailing exposures to high doses and deterministic-type effects, the field of radiation protection addresses public's and workers' exposure to low doses for which the ensuing problems are limited to the occurrence of stochastic phenomena.

Determination of the risk associated with such exposures requires the definition of the dose-frequency relationship for each of these damages.

This dose-frequency relationship can be derived from in vitro studies performed on cell cultures, from animal experiments and from epidemiological investigations.

In practice, monitoring exposed human populations is the only valid basis for defining the risk factor.

Cancers are responsible for approximately 30% of the overall mortality in Western countries, the second leading cause of death after cardiovascular diseases. Therefore, excess cancers developed by a population exposed to ionising radiation can only be assessed by comparison with a control population having the same characteristics of age, sex and exposure to other factors (tobacco, alcohol, food, environment... the suspected causes of 60-80% of human cancers). This is the purpose of epidemiological surveys.

In addition to the difficulty of defining a suitable control population, epidemiological investigations face a number of methodological problems:

- effects are non-specific, and because cancer-related mortality in Western populations is approximately 30%, excess cancers due to exposure to low doses are difficult to substantiate;

- the lower the exposure dose, the lower the number of potentially radiation-induced cancers will be, and the larger the study population should be. Indeed, it is recognised that it is necessary to work with cohorts of about 1000 people for an exposure of 1 Gy, 10 000 people for an exposure of 0.1 Gy, and 10 000 000 people for an exposure of 0.01 Gy;

- because the effects are delayed, the study calls for a life-long monitoring.

In practice, the population of survivors of the Hiroshima and Nagasaki bombings was chosen for risk calculations. The study includes approximately 86 000 subjects.

A significant excess of leukaemias and cancers was observed among the subjects exposed to more than 0.2 Gy.

Below 0.1 Gy, there is no reliable data which allows an estimate of a carcinogenic effect in adults.

To determine a dose-response relationship for doses below 0.1 Gy, defined in this context as the low doses domain, the data obtained at higher doses should be extrapolated, although this method involves a number of uncertainties.

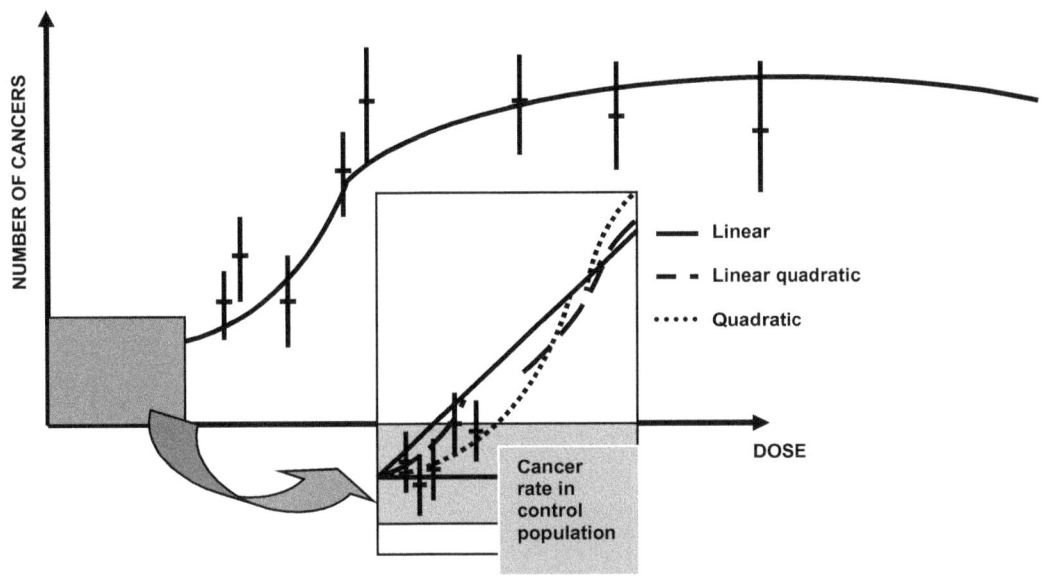

Figure 4.10. Dose-effect relationship.

Figure 4.10 represents the dose-effect relationship established from cancers observed among survivors of the Hiroshima and Nagasaki bombings, and it exhibits several domains:

- doses > 1 Gy: the number of excess cancers is significant, the uncertainty is important, but the shape of the curve would be of linear quadratic type;
- higher doses: the linear quadratic relation bends due to cell death;
- doses < 0.1 Gy: data is scarce and the low number of excess cancers is not significant compared to the value in the control population.

Different mathematical expressions consistent with the data are shown in the inset of Figure 4.10.

Extrapolation to low doses

In the low doses domain, it is necessary to extrapolate the dose-response relationship that was determined for exposure to high doses.

To this end, the precautionary principle has prevailed and the most pessimistic assumptions were retained, leading to a linear extrapolation going through the origin:

- any dose is associated with an additional risk; there is no threshold;
- the risk is proportional to the dose.

These concepts are summarised by the commonly used term "Linear No-Threshold" relation.

The fundamental practical conclusion is to recognise that any exposure, even the slightest, is associated with a risk; its corollary is to accept the risk if one accepts the exposure. This case is very different from that of deterministic effects, for which the risk can be prevented by simply observing the exposure threshold.

4.6.2. Genetic effects

The largest survey, covering 30 000 children from Hiroshima and Nagasaki, each with at least one parent irradiated during the bombing, showed no significant increase in the frequency of genetic anomalies.

Therefore, estimates of the genetic risk must be extrapolated from experimental data obtained with insects or, for relatively high doses, mice.

Although the linear quadratic relation fits the experimental facts, human extrapolations rely on the LNT relation used in carcinogenesis.

4.6.3. Quantification of the total risk of stochastic effects

The linear no-threshold extrapolation allows international commissions to estimate an excess risk after exposure to radiation in the low doses domain. The total risk is the sum of a somatic risk (carcinogenic) and a genetic risk. The numbers currently in use date back to the ICRP Publication 60 and are the basis for current regulations.

4.6.3.1. Carcinogenic risk

The probability coefficient regarding fatal and non-fatal cancers is $4.80 \cdot 10^{-2}$ Sv^{-1} for workers.

4.6.3.2. Genetic risk

Still in the case of a whole body irradiation, the probability coefficient regarding severe hereditary effects is $0.80 \cdot 10^{-2}$ Sv^{-1} for workers.

4.6.3.3. Total risk

Therefore, the probability coefficient regarding all stochastic effects is $5.6 \cdot 10^{-2}$ Sv^{-1} for workers.

4.6.4. The concept of radiation detriment

It is with the ICRP Publication 60, issued in 1991, that the concept of radiation detriment evolved. Previous recommendations (ICRP 26, 1977) were based on a characterisation of the health detriment in terms of risk of fatal cancers and hereditary anomalies in the first two generations. With ICRP 60, radiation detriment represents the risk of cancer adjusted for lethality and quality of life, and the risk of hereditary effects adjusted for lethality and quality of life. Thus, it considers four points:

- the incidence of fatal cancers (number of fatal radiation-induced cancers according to organ) and non-fatal cancers;
- the weighting factor for quality of life, which applies to the non-fatal cancers. The ICRP considers it necessary to include the pain, suffering and side effects related to cancer treatment into this weighting factor;
- the risk of hereditary anomalies for future generations;
- the relative loss of life expectancy in cases of fatal cancer or serious hereditary disorders.

The latest recommendations for the management of radiation risk rely on ICRP 103

As previously mentioned, current radiation protection regulations are based on general recommendations for human and environment protection against ionising radiation, originating with the 1991 ICRP Publication 60. In 2007 the ICRP issued new recommendations which were the subject of publication 103. Here, the commission relied on a foundation of scientific knowledge derived from studies carried out in the mean time on the biological effects of exposure to ionising radiation, in particular on data from UNSCEAR publications. The ICRP concluded that, after their extensive review of the literature published after 1991, there were no fundamental changes which suggested the need to modify the radiation protection system.

Regarding the risk of radiation-induced cancer, the shape of the dose-effect relationship is maintained. Indeed, the increase in the number of observed cancers contributes to the accuracy of this relationship; moreover, the data obtained from the monitoring of the Japanese survivors of Hiroshima and Nagasaki (which, to reiterate, represents the largest cohort studied for an extended length of time with a realistic assessment of individual doses) is compatible with a linear or linear-quadratic relation in a wide range of doses. Regarding the issue of threshold, the dose below which exposure would not induce cancer, the ICRP considers that the system of low dose protection can continue to be based on a linear no-threshold extrapolation. With this the ICPR adopts the conclusions of the UNSCEAR that, although epidemiology alone is unable to demonstrate the existence or non-existence of a threshold dose below which radiation would not induce cancer, this does not mean that such a risk does not exist. The dose and dose rate effectiveness factor (DDREF) is maintained at a value of 2 as set in 1990. Finally, regarding the quantification of the cancer risk, the coefficient of total radiation detriment associated with exposure to ionising radiation is estimated at $5.5 \cdot 10^{-2}$ Sv^{-1} for the entire population, and $4.1 \cdot 10^{-2}$ Sv^{-1} for workers aged 18 to 64 years, a level of risk of the same order of magnitude as that recommended in 1990.

Regarding the risk of radiation-induced hereditary disease (genetic effect), the ICRP has considered new data on the quantitative aspects of genetic mutations expressed in successive generations. The commission now compares the rate of spontaneous mutations in human genes and the rate of radiation-induced mutations in mouse genes, whereas the previous estimate was based solely on the mouse, leading to a risk overestimation. The current estimate for the risk of hereditary diseases until the second generation is $0.2 \cdot 10^{-2}$ Sv^{-1}.

Therefore, the 1990 and 2007 values for the global risk of detriment due to radiation-induced cancer and hereditary effects differ little for cancers, whereas hereditary effects are reduced by a factor of 6 to 8. To establish international standards on radiation protection, the ICRP recommends a value of approximately $5 \cdot 10^{-2}$ Sv^{-1} for the global detriment.

Further information

Definition of the term "incidence" (cancer incidence)
Fédération Nationale des Centres de Lutte Contre le Cancer (National Federation of French Cancer Centres)
Bernard Hœrni, 16/5/2002 – updated on: 15/12/2005

In epidemiology, the incidence is the number of new cases of a disease observed during a given period and for a given population; it is the main criterion for assessing the frequency of cancers. It differs from the prevalence which includes all actual cases at a given time. Cancer incidence is usually expressed for a year, as an absolute value for a country or region, or as a relative ratio per 100 000 inhabitants. Its value can be adjusted according to the age profile of a population in order to allow international comparisons. Without this adjustment, it would be difficult to compare developed and developing countries, whose age profiles differ greatly, due to the dominating influence of aging on the occurrence of cancer.

For most countries, an increase in cancer incidence can be observed, but most of this increase is due to the aging population and vanishes when you consider people of the same age, for example the age range between 60 and 70 years old. Another part of the increase is due to cancer being diagnosed more often, whereas previously, deaths were not attributed to any specific cause (other than "old age"); screenings, which continue to develop, also lead to identifying cancers earlier than they would have been by chance and a few cancers which would not have been detected until after the person's death. Finally, the slight increase in incidence observed in industrialised countries is primarily due to a sharp increase in smoking-related cancers, first and foremost bronchial cancer, which conceals a slight decrease in the incidence of all other tumours.

Definition of the term "mortality" (cancer mortality)
Fédération Nationale des Centres de Lutte Contre le Cancer (National Federation of French Cancer Centres)
Bernard Hœrni, 16/5/2002 – updated on: 24/02/2003

Mortality expresses the rate of death per unit of time for a given population, usually per year per 100000 inhabitants. It varies with the age of the population. This is especially true for cancers whose frequency increases with age (only 1% of cancer deaths occur before the age of 35). Thus, cancer mortality is higher in industrialised countries, where the population includes a higher proportion of elderly than in developing regions where the population is younger (and dies earlier from other causes). [...]

[...]Worldwide about 50 million of people die each year. One tenth of these deaths (5 to 6 million) are linked to cancer, but this number is expected to rise to 12 million in 2020. By comparison, 17.5 million deaths are caused by infectious and parasitic diseases,

12 million by cardiovascular diseases, and 500000 by pregnancies. As cancers occur mainly in the elderly – in contrast with infectious diseases which kill many children in the third world – their impact is diminished if we take the number of years of life lost as a reference: worldwide, cancer is responsible for about 6% of these losses, versus 45% due to infectious diseases; to show the two extremes, these numbers are respectively 19% and 10% for industrialised countries, and 2% and 71% for sub-Saharan Africa. [...]

[...] In most countries, besides bronchial cancer, there is an increase in mortality from breast, prostate and colon cancers. By contrast, mortality from stomach and cervix cancers decreased noticeably. An evaluation was conducted on the part played by different cancer-causing factors in mortality due to malignant tumours in Western countries. Nutrition appears to be the cause of one third of cancer deaths, closely followed by tobacco (30%). Other factors come far behind: sexuality (7%), occupation (3-4%), alcohol (3%), industrial products (1%). A number of cancers remain without any identified cause. Sun exposure, causing many tumours, plays only a modest part because most skin cancers are healed.

Carcinogenesis and the notion of associated probability
(Clefs CEA, 2000)

In humans, the cells of a healthy organism can only divide at the request of the surrounding cells. By contrast, cancer cells ignore the signals that would limit their proliferation, and obey only their own duplication program. The cells of a tumour are all produced by the abnormal division of a single cell at the origin of a tumoral clone. The process, which spreads over years or decades, results from an accumulation of mutations in certain genes of this cell.

Two classes of genes, protooncogenes and tumour suppressor genes, encode proteins involved in the regulation of the cell cycle, and play a fundamental role in the occurrence of cancer. Protooncogenes stimulate cell growth, while tumour suppressor genes inhibit it. The loss or deletion (caused by breaks in both DNA strands due to ionising radiation or another mutagenic agent) of a chromosome fragment carrying a tumour suppressor gene may inhibit mechanisms for stopping cell division.

Genes encoding repair systems are also involved in carcinogenesis. The introduction of mutations in these genes results in DNA instability, since all cells of the tumour contain DNA that was repaired variably and thus differs from one cell to another.

DNA damage, activation of an oncogene or inactivation of a tumour suppressor gene normally cause cell suicide by apoptosis, which constitutes a defence against the proliferation of potentially cancerous cells.

A second line of cell defence against proliferation involves the shortening, at each DNA replication, of DNA segments located at the ends of the chromosomes (telomeres), which record the number of cell divisions. When they become too short, these telomeres no longer protect the chromosome ends, resulting in many fusions between chromosome ends. This generates unstable chromosomes and a genetic chaos fatal to the cell. Telomerase, the enzyme that replaces the telomere segments removed at each division, is absent from almost all healthy cells but present in almost all tumour cells.

Mutations in genes related to apoptosis and in the gene encoding telomerase may also be involved in the carcinogenesis process.

As a result, a tumour develops in stages and occurs after the mutation of several genes, which explains the long period that generally passes between the original mutation and the clonal evolution of the cell which hosted the original mutation.

4.7. The principles of the ICRP

Established in 1958, the ICRP is a non-governmental organisation composed of independent international experts with the aim of issuing recommendations.

Accordingly, its actions revolve around three main principles:

- **justification of practices**: no practice involving exposure to ionising radiation should be adopted if it does not provide an ADVANTAGE that outweighs the DETRIMENT it may cause;

- **optimisation of exposure**: protection optimisation is based on the adoption, on principle, of the LNT dose-effect relationship. Appreciating that any exposure will generate an effect, the ICRP requests that any occupational exposure be kept to the lowest possible dose level, taking into account economic and social factors;

- **limitation of exposure doses**: for deterministic effects with a threshold, the limit is set below the threshold; for random effects without a threshold, the limit is set at a level of acceptable risk.

Further information

Recommendation of limits by the ICRP (Table 4.5)

To determine these values, the ICRP studied the life expectancy of a population exposed to ionising radiation. It compared the effects of an exposure reaching the limit value during each year of the professional life of a worker, to the life expectancy curves of control populations. Given the risk factors and taking into account the latencies of the various cancers that can result from exposure to ionising radiation, the commission assessed the loss of life expectancy for different dose levels ranging from 50 mSv/year to 20 mSv/year. The ICRP concluded that the value of 20 mSv/year does not reduce the life expectancy of a worker exposed to this level of radiation for his/her entire professional life. The commission considers that the risk linked to an exposure of 20 mSv/year is acceptable, and this limit is recommended to regulate exposure to ionising radiation. Despite this allowance, and following the optimisation principle, the ICRP recommends that regulations and workplaces aim for the lowest possible exposure.

Table 4.4. New values of the weighting factors for tissues, following the recommendations of ICRP 103 published in 2007.

Tissue or organ	Weighting factor w_T for tissues New recommendations
Gonads	0.08
Bone marrow (red)	0.12
Colon	0.12
Lung	0.12
Stomach	0.12
Bladder	0.04
Breast	0.12
Live	0.04
Oesophagus	0.04
Thyroid	0.04
Skin	0.01
Bone surface	0.01
Brain	0.01
Salivary glands	0.01
Other tissues or organs	0.12

Table 4.5. Recommended limit values (ICRP, Publication 103).

Workers	**Effective dose**	20 mSv/year on average over 5 years and not exceeding 50 mSv/year
	Pregnant women	< 1 mSv on the surface of the abdomen/9 months
	Equivalent dose Crystalline lens	150 mSv/year
	Skin, hands, feet, ankles	500 mSv/cm^2/year to a depth of 7 mg/cm^2
General public	**Effective dose**	1 mSv/year on average over 5 years
	Equivalent dose Crystalline lens	15 mSv/year
	Skin, hands, feet, ankles	50 mSv/year

4.8. Check your Knowledge

You can now check the knowledge you have acquired from this chapter by answering the following questions:

1. **Fill in the blanks in the following text (one or two word(s) per blank space):**

    Radiation interacts by _____ when the energy is directly transferred to the molecule of biological interest, namely DNA.

    The _____ effect consists of the modification of the molecule of biological interest, namely DNA, through chemical reactions with the _____ formed after ionisation of a water molecule due to the radiation's interaction.

    Answer: direct effect, indirect, free radicals.

2. **Which of the following cellular organelles has been identified as a critical target for ionising radiation:**

    a. cytoplasm

    b. water

    c. lysosome

    d. DNA

    Answer: d.

3. **When the DNA of a cell is altered by interaction with radiation, what are the three possibilities for the future of this cell?**

    Answer:

    – Faithful repair of the induced damage, the cell regains its normal gene composition.
    – Incorrect or incomplete repair leading to a "mutant" cell.
    – Cell death.

4. **Deterministic effects are characterised by:**

    a. the existence of a threshold dose

    b. no effect below the threshold dose

    c. above the threshold dose, the severity of the effects increases with the dose

    d. all of the above

    Answer: d.

5. **An effect whose probability of occurrence increases with the dose is called:**

    a. stochastic effect

    b. deterministic effect

    c. non-stochastic effect

d. all of the above

Answer: a.

6. **Which types of biological effects are likely to occur at low doses for humans:**

    a. deterministic effects

    b. early-onset effects

    c. hereditary effects

    d. stochastic effects

    Answer: d.

7. **Among the following effects, which are stochastic effects:**

    a. carcinogenesis

    b. hereditary effects

    c. cataract

    Answer: a. and b.

8. **Epidemiological studies have demonstrated a significant risk of cancer for exposures above:**

    a. 10 mSv

    b. 100 mSv

    c. 1000 mSv

    d. 2000 mSv

    Answer: b.

9. **Which hypothesis was chosen on the mathematical form of the dose/excess cancer curve for extrapolation to low doses:**

    a. quadratic

    b. LNT

    c. linear-quadratic

    d. exponential

    Answer: b.

10. **The role of radiation protection is to:**

    a. prevent the occurrence of stochastic effects and reduce the occurrence of deterministic effects

    b. reduce the occurrence of deterministic effects only

    c. prevent the occurrence of deterministic effects and reduce the occurrence of stochastic effects

    d. prevent the occurrence of deterministic effects only

    Answer: c.

11. **Do the following sentences refer to a deterministic or stochastic effect?**

    a. The effect that is likely to occur after a whole-body exposure to low doses is _____

    b. What kind of effect is likely to occur after a whole-body exposure to high doses received over a short time period? _____

    c. An effect exhibiting a threshold dose is _____

    d. For what type of effect is the severity independent from the dose received? _____

    Answers: stochastic, deterministic, deterministic, stochastic.

12. **With the help of Table 3.5, order the following organs and tissues from the least radiosensitive to the most radiosensitive: thyroid, breasts, gonads, skin, bone marrow.**

    Answer: skin, thyroid/breasts, bone marrow, gonads.

13. **List the characteristics of deterministic and stochastic effects, respectively.**

    Answer:

    Deterministic effects:

    - existence of an exposure threshold above which lesions or pathologies will be observed in all subjects (hence the term deterministic);
    - the severity of damage increases with the dose;
    - early onset (days, months), except for cataract.

    Stochastic effects:

    - occur randomly, only in some individuals;
    - the probability of occurrence increases with dose;
    - the severity is independent of the dose, effects are all or nothing;
    - late onset: 5 to 10 years for leukaemia, 20 to 50 years for most other cancers, at least a generation for genetic effects.

14. **What are the major principles of radiation protection? Explain them.**

    Answer:

    - The justification of practices: no practice involving exposure to ionising radiation should be adopted unless it provides an ADVANTAGE that outweighs the DETRIMENT it may cause.
    - The optimisation of exposure: any exposure shall be kept at the lowest possible level, taking into account economic and social factors.
    - The limitation of exposure doses: deterministic effects with threshold: limit set below the threshold; random effects without threshold: limit set at an acceptable risk level.

5 Detection and measurement of ionising radiation

Jean-Christophe Bodineau, Marc Ammerich, Jean-Claude Moreau, Abdel-Mjid Nourreddine

Introduction

Detection of radiation is based on an interaction process between the radiation and a medium capable of delivering a signal proportional to the interaction. This medium is known as the "detector".

"Active" detectors allow the immediate measurement of any interacting radiation. These detectors can provide three types of information:

- either a "count" representing the interaction between the radiation and the medium. The accumulation of detection counts is called a "counting";

- or by additionally providing an indication of the amount of energy transferred from the radiation to the detector; when this additional information is used, the process is then referred to as "energy spectrometry";

- or by delivering an average current representing the flux of radiation interacting with the detector, in which case it is no longer possible to count the individual interactions.

In some detectors, the signal cannot be extracted immediately after the interaction. Instead, the signal corresponding to a radiation flux is generated only well after the exposure period, integrated over this period of time. These detectors are called "passive" detectors.

The energy that is transferred to the detector medium often triggers both ionisations and excitations. The proportion between both phenomena is, on average, constant for a given medium. In a detector medium, both phenomena occur, but it is seldom possible to simultaneously extract both types of signals. Indeed, some mediums are opaque, and the light resulting from de-excitation cannot escape, while other mediums do not conduct electricity, making it impossible to collect the charges created by ionisation. It should be noted that the majority of mediums are not suitable as detectors.

Heat can be measured after energy deposition by radiation in a medium. But at the level of doses measured in most common situations, heat cannot be used for detection because the energies involved remain very low. However, at the scale of particle accelerator beams, this effect is measurable by using calorimeters.

The first section of this chapter describes the operating principles of detectors.

The second section covers the electronic devices which provide a "measurement result" from the "primary signal" produced by a detector; these devices as a whole constitute the detection chain.

The measurement methods are the subject of the third section. It first describes active detectors used to count interactions triggered by radiation, as well as the measurement of average ionisation currents in order to measure either activities or dose rates. Then the second part of this section introduces passive detectors used to integrate the absorbed doses.

5.1. Detectors

5.1.1. Scintillation counters

Historically, scintillators have long been limited to detection of radiation by fluorescent screens with optical analysis of the radiation-generated scintillation. X-rays, have been discovered thanks to fluorescent screen that was a scintillator allowing the visualisation of relatively intense fluxes of radiation. The scintillation caused by a single radiation could only be perceived in total darkness, with the help of a large optical magnification and only for particle with high linear energy transfer such as α radiation.

The use of these detectors increased with the appearance of photomultiplier tubes.

5.1.1.1. Physical phenomenon exploited: excitation and de-excitation

When ionising radiation interacts with matter, the energy of the radiation is transferred to the atomic electrons due to 'collisions' with these electrons. When the energy transferred is larger than the electron binding energy, the electron is ejected from the atom leading to ionisation. When the tranferred energy is lower than the electron binding energy, the electron does not leave the atomic structure, but moves to a higher electron energy level: this process is called "excitation". The electron can stay at this higher electron energy level for various times, but ultimately return to its intial energy level: this process is called "de-excitation". This return to the initial energy level is accompanied by the emission of light: this is called "luminescence". The phenomenon is further called "fluorescence" when the de-excitation occurs at short times after excitation, and "phosphorescence" when luminescence occurs less rapidly after excitation. For more information, please refer to Chapter 2, "Interaction of ionising radiation with matter".

5.1.1.2. Overview and structure of a scintillator

The scintillator and other constituent parts of a typical scintillation counter are presented in Figures 5.1 and 5.2.

Scintillators use the fraction of energy that is dissipated in atomic or molecular excitations and de-excitations. If the dimensions and specifications of the scintillator allow for the total dissipation of the incident radiation's energy, then it is possible not only to detect this radiation, but also to measure its energy.

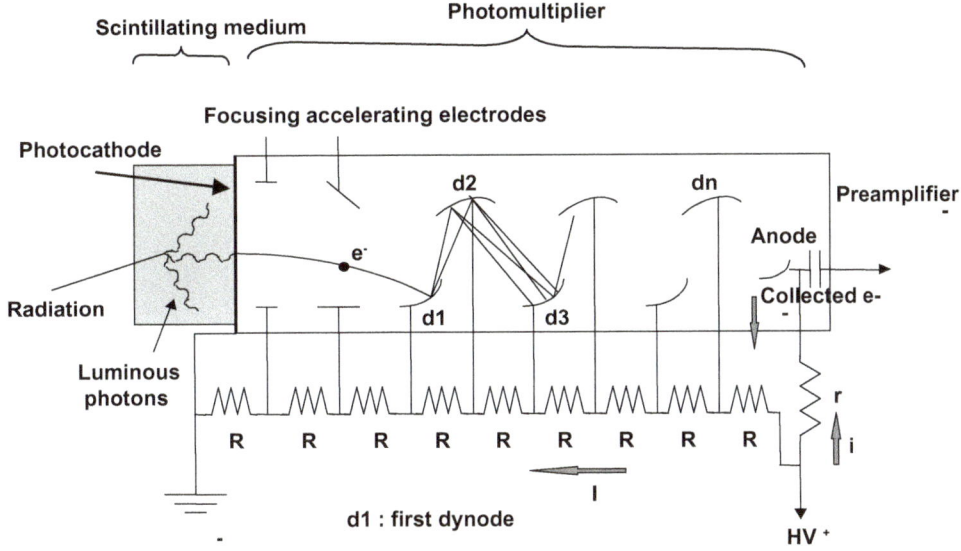

Figure 5.1. Structure of a scintillation counter.

Figure 5.2. Structure of a scintillation counter (source INSTN).

Different materials are used:

– mineral scintillators: thallium doped cesium iodide (CsI) and sodium iodide (NaI) for the detection of X and gamma radiation; zinc sulphide (ZnS) for the detection of alpha radiation;

– organic scintillators, in solid form (anthracene, plastics) for the detection of any type of ionising radiation or in liquid form (particularly for the detection of low-energy β radiation).

The scintillation efficiency, which corresponds to the fraction of incident energy that is converted to light, is low: 4 to 8% depending on the type of scintillator. Luminescence photons are mainly in the blue range, with an extension of the spectrum in the near UV.

The scintillation light (all photons that are emitted due to the interaction process in the scintillator) decreases over time according to an exponential law whose time constant varies between 2 and 30 ns for organic scintillators, and can reach approximately 1 µs for mineral scintillators. The most important characteristic of a scintillator is that it must be transparent to its own light, e.g. the flash of photons formed in the scintillator must be able to exit the scintillator in order to be converted into electron pulses in the photomultiplier. Indeed, the "light signal" cannot be used unless it is transformed into an "electric signal" of high amplitude. The scintillator is therefore coupled to a photomultiplier which converts the flash of light into electrons in the photocathode. The number of electrons produced is then significantly increased thanks to a series of dynodes.

In common organic scintillators, one luminous photon is generated for every 30 eV of ionising radiation energy that is transferred to the detector. For some very special cases of liquid scintillators, the low energy of the incident radiation does not allow for any interposition between the source and the scintillator. Tritium is a perfect example, because with an $E_{\beta max}$ of 18.6 keV, its maximum range in water is 7 µm.

To conduct a measurement, it is necessary to thoroughly mix the source with a liquid scintillator. The latter consists of aromatic molecules belonging to the solvent family.

One phenomenon to take into account is the dark current of photomultipliers: due to variations in temperature, electrons in the photomultiplier can be spontaneously and randomly removed from the photocathode resulting in a signal that is not related to the ionising radiation. In order to precisely distinguish pulses resulting from the interaction of a particle with the detector from 'false' pulses, the scintillator is coupled to two photomultipliers which are in turn coupled to a coincidence circuit which transmits the signal only if both input ports receive two simultaneous pulses. Such a device, presented in Figure 5.3, thus eliminates the counting of pulses due to either photomultiplier's background.

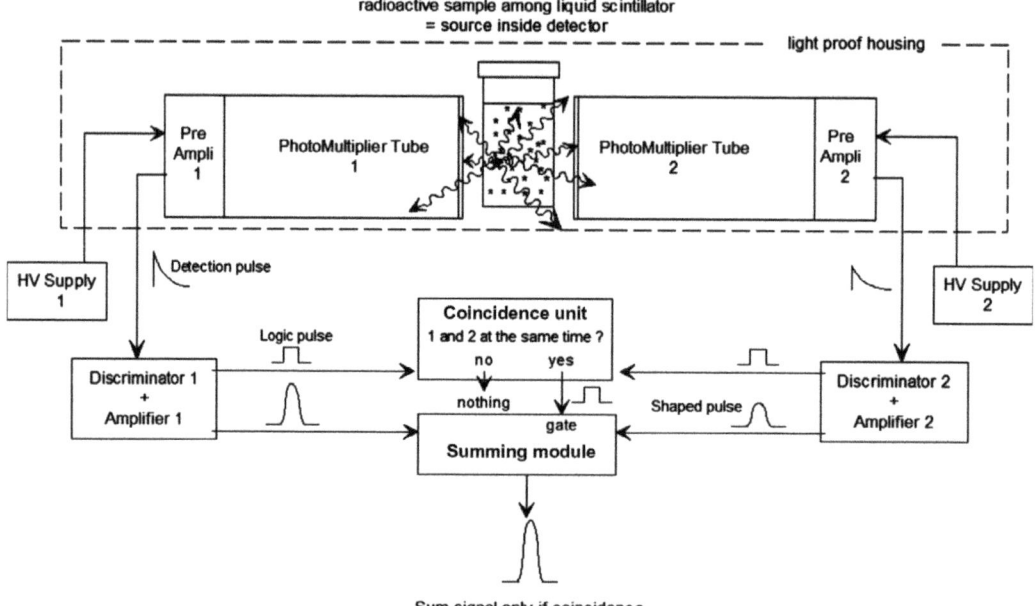

Figure 5.3. Coincidence measurement of detection signals with liquid scintillation.

In liquid scintillation, the counting efficiency depends on the measured radionuclide as well as on the dimming phenomenon called "quenching". This phenomenon depends on the colour and the nature of the chemicals introduced in the scintillating medium, whose luminescence properties are more or less degraded from one sample.

The amount of luminous photons produced by scintillation is converted into a very small number of electrons obtained by photoelectric effect in a photocathode thin enough to allow the photo-electrons to escape.

5.1.1.3. Operation of a photomultiplier

Since the production and collection efficiencies of luminous photons, as well as the conversion efficiency into electrons, are relatively low, the "electronic signal" must be amplified in order to be usable: this is the role of the electron multiplier, which allows for an electronic pulse of high intensity as an output, thanks to a series of dynodes.

The semi-transparent photocathode has an efficiency which depends on the incident light's wavelength. The quantum efficiency, e.g. the number of electrons emitted per incident luminous photon, is at maximum 25%.

The electrons emitted by the photocathode are accelerated and directed to the first dynode by a potential gradient through a set of electrodes.

Upon arriving at the first dynode, each electron has sufficient kinetic energy to remove more electrons from the metal surface.

The secondary electrons, obtained from the first dynode, are guided and accelerated by an electric field to a second dynode where each remove in turn more electrons, and so on. A device with n dynodes has a multiplying coefficient $A = \delta^n$ and a large number of electrons are finally collected on the anode at the end of the electron multiplication tunnel.

The δ factor is called the secondary emission coefficient and varies between 4 and 5 when the voltage between stages is of the order of 150 volts.

The voltage between each dynode is created by a high voltage supply applied to the tube and shared with all dynodes using a series of resistors. The high voltage applied to a photomultiplier tube ranges from 500 V to 1500 V according to the tube and the intensity of the scintillations to be converted.

The quantity of electrons collected on the anode is proportional to the amount of scintillation light. If this light itself is proportional to the energy loss incurred by the incident particle in the scintillator, then by measuring the amplitude of the electronic pulse, we have an indication of the amount of energy deposited by the radiation and a "spectrometry of the radiation's energies" becomes possible.

The high voltage supply of the photomultiplier must be very stable, because even a small change in voltage causes a large variation of the electron multiplication gain created in the tube. When scintillators are used for energy spectrometry, the spectra frequently require recalibration by modifying the gain of the measurement chain's amplifier.

Furthermore, when photomultipliers, coupled with their scintillators, are used for energy spectrometry, large fluctuations in the number of electrons generated by the same scintillation lead to wide peaks in the spectra, enabling only low energy resolution. This does not allow separation of the information resulting from several radionuclides mixed within the same sample.

5.1.1.4. *Implementation of the scintillator-photomultiplier coupling*

Because the scintillation flashes are so weak, it is imperative that the "scintillator + photomultiplier" set is protected from external light.

When scintillators (for example, NaI) are sensitive to humidity, they are packaged in a sealed assembly with only one side equipped with a glass window designed to be coupled to the photomultiplier.

To ensure proper collection of the luminous photons, the scintillators are partially surrounded by a reflector which can act as a seal from the light.

The optical coupling between the scintillator and the photocathode is obtained through a medium with an intermediate refractive index to reduce light reflections. Grease, transparent glue and silicone paste are the most commonly used media. When necessary, light guides can be mounted between the detector and the photomultiplier.

The ideal scintillator should exhibit the following properties:

- high efficiency in converting kinetic energy deposited by the radiation into photons;

- linear conversion on a wide range of energies;

- high transparency of the medium in the range of the wavelength of the emitted photons;

- very short lifetime of the induced luminescence (to avoid signals that last too long, are too weak and do not allow the counting of events that are clearly distinct in time);

- material which can be implemented on a large scale and which is easy to manufacture particularly for the production of portable and well-shaped detectors;

- high atomic number and high density for the detection of gamma radiation;

- light emitted in the visible range to avoid the production of sophisticated photomultipliers;

- refractive index close to the glass index to facilitate coupling with the photomultiplier.

The scintillators most commonly used for radiation protection are:

- mineral scintillators NaI (NaI(Tl)), LaBr ($LaBr_3(Ce)$) for the measurement of gamma radiation and for spectrometry;

- mineral scintillator ZnS (Ag) for the counting of alpha radiation;

- solid organic scintillators (plastics) for the counting of β and γ radiation;

- liquid organic scintillators for the spectrometry of β radiation, mostly of low energy, or even α radiation.

5.1.2. Gas-filled detectors

5.1.2.1. Physical phenomenon exploited: ionisation

When crossing a medium, charged particles lose energy in collisions with nuclei or electrons of the medium's atoms. In the energy range of the particles emitted during radioactive decay, as seen in Chapter 2, "Interaction of Radiation with Matter", collisions with the atomic electrons are the dominant interaction process.

When the energy released by the incident particle is greater than the electron's binding energy, the electron is ejected and consequently, a positive ion is created. In this way, a primary electron-ion pair, also called "ion pair", appears: this phenomenon is known as ionisation.

Positive ions eventually recombine with either the medium's free electrons or in the vicinity of electrodes connected to a voltage and current supply.

5.1.2.2. Overview of a gas-filled detector

A gas-filled detector, or counter, is made of a gas- (or gaseous mixture-) filled chamber and two metal electrodes between which a potential difference V is applied. V depends on the distance between the electrodes and is often a high value applied through an HV (High Voltage) supply. A scheme of a gas-filled detector is shown in Figure 5.4. Cylindrical configurations are the most commonly encountered for radiation protection.

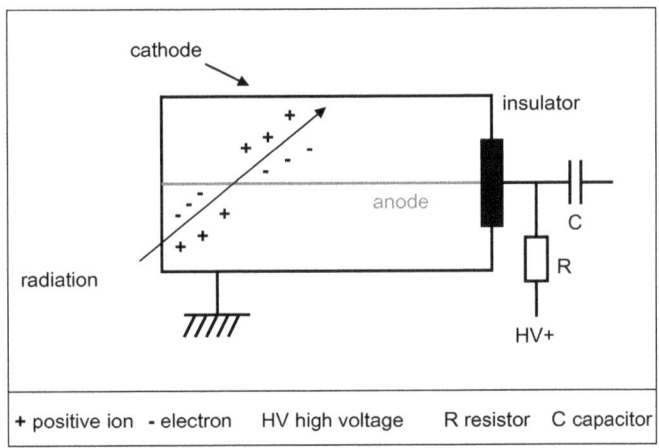

Figure 5.4. Schematics of a gas-filled detector.

As described earlier, when radiation passes through the gas, it is ionised and many ion pairs (negative electron and positive ionised atom) are created. In the absence of voltage between the two electrodes, the ions formed, which are subject to thermal agitation, quickly recombine to again form neutral atoms. This is called initial recombination.

When the particles interact with a gas that has been subjected to a static electric field obtained by application of a continuous voltage between two electrodes, the positive ions are attracted by the cathode (−) and the electrons by the anode (+). The electrodes collect charges which are able to flow into a resistor. This results in an ionisation current with an intensity proportional to the total number of ion pairs created. Therefore, a voltage or current pulse corresponds to any ionising radiation which has interacted into the gas. This is the mechanism upon which the operating principles of gas-filled detectors rely.

Because of the mass difference between electrons and positive ions, the velocity of the electrons is much higher than that of the ions (about 10 000 times larger).Therefore it is preferable to use the signal from electron collection in order to shorten the time between the interaction of the radiation in the gas and the occurrence of a detection pulse.

On its trajectory, a particle creates on average a number of ion pairs equal to:

$$N_{pi} = \frac{E_0}{(\varpi)}$$

with
E_0: energy transferred by the incident radiation to the gas;
ϖ: energy that radiation must, on average, spend to create an ion pair.

For gases, ϖ is on the order of 30 to 35 eV and is nearly independent from the nature of the incident radiation and from its energy; ϖ depends weakly on the filling gas.

If the detector had an infinite capacity C_d, the charges created by ionisation would accumulate in the capacitor and it would be impossible to measure a voltage pulse. C_d being finite, it discharges through the resistor R following the time constant: $\tau = RC_d$. If the value of τ is larger than the electronic collection time and smaller than the ion collection time, only the current signal due to electrons can be detected. The result is a voltage variation across the resistor R, which can be measured behind the capacitor C.

The best choice for a filling gas is one with no affinity for electrons. This avoids the loss of electrons in the signal due to their recombination in the gas itself.

Radiation can interact not only with the detector's gas but also with its internal walls, thus producing secondary particles which are detected by the gas. This "wall effect" is usually undesirable but can be useful, especially to individually detect high energy gamma photons that have a very low probability of interaction in gases.

5.1.2.3. Operating regimes

Figure 5.5 shows the variation in the number N of charges collected on the electrodes as a function of the potential difference V_{HT} applied to polarise the detector, $N = f(V)$. However, the full range of this figure does not apply to a single gas and a single detector.

The two curves represent the number of ion pairs N_1 created with an energy E_1 and the number of ion pairs N_2 with an energy E_2, respectively, as a function of the applied voltage.

The curves indicate five distinct regions which reflect different operating modes:

a) Region I: the electric field applied between the electrodes is too weak to prevent all or part of the ion pairs from recombining. This leads to an insufficient number of charges collected on the electrodes. No detector operates in this region.

b) Region II: all of the electrons and ions created are collected. The number of charges collected is equal to the number of charges produced by primary ionisation: $N_c = N_{pi}$. There is no amplification of the number of charges created because the electrons are not

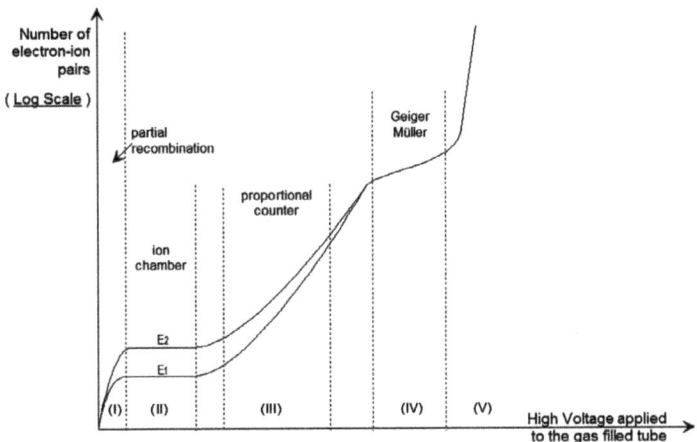

Figure 5.5. Different operating modes of a gas-filled detector as a function of the high voltage applied for events depositing two different amounts of energy within the gas.

sufficiently accelerated by the electric field to ionise the gas: there is no secondary ionisation. This is the operating mode of "**ionisation chambers**".

Even when the number of electrons collected is proportional to the energy transferred by radiation in the gas, ionisation chambers are rarely used for spectrometry. Their application in this field is limited to the grid chamber, which is used for α radiation spectrometry, which can be found in some laboratories performing radioactivity analysis.

Likewise, they are seldom used to perform interaction counts, except for some fission chambers used to measure neutron flux in and around nuclear reactors.

In radiation protection, they are instead used to measure photon and electron radiation fluxes, where, as we shall see later, their detection signals deliver an average current whose intensity is proportional to the absorbed dose rate.

In practice, the operating voltage varies between 60 and 300 V depending on the dimensions of the chamber.

c) Region III: the electric field is strong enough to convey sufficient energy to the primary electrons to allow them, in turn, to ionise the gas atoms. This multiplication process takes the form of a avalanche, known as an "townsend avalanche" in which each free electron created in turn generates yet more free electrons through the same process. The number of electrons finally collected is proportional to the number of primary electrons: $N_c = k.N_{pi}$, k being the multiplication coefficient or amplification factor.

This region corresponds to the "**proportional counters**" operating mode. In this region, the multiplication coefficient in the gas is a function of the voltage only; it increases very quickly as the voltage increases (it can reach 10^6).

Besides the counting of radiation interactions, proportional counters can also be used for energy spectrometry (for example, low energy X-rays or electrons), provided that the high voltage (HV) is well stabilised, since a small variation of V results in a large variation of k, as in photomultipliers.

The proportionality between the amplitude of the electronic detection signal and the energy released by the radiation is often used to distinguish interactions due to α particles from those due to β radiation. This is utilised in "hand and foot monitors", skin and even clothing contamination monitors, which we will discuss later.

In practice, the operating voltage varies between 500 and 4000 V, depending on the size and on the filling gas. The nature of the gas and its pressure vary greatly depending on the application. In gas mixtures, noble gases (Ar, Kr, Xe...) are preferred over molecular gas, which tend to recapture electrons and to reduce the gas amplification. In practice, mixtures such as "Ar + CO_2" or "Ar + CH_4" are often used.

For neutron detection, proportional counters are widely used when a "converter" such as ^{1}H, ^{10}B, ^{3}He, or even ^{6}Li is used. The converter allows, through nuclear reaction, the transformation of the neutron into heavy, directly ionising particle(s). It can be found either in the constituents of the detector gas, for example ^{3}He ou ^{10}BF$_3$, or distributed as a thin layer on the inner surface of the cathode (for example boron lined counters, enriched in ^{10}B).

d) Region IV: the importance of the ion multiplication is such that the number of electrons collected is independent from the number of primary ions created by the radiation which interacted with the gas. This generates an almost complete ionisation of the gas volume surrounding the anode, provided that the radiation has created at least one ion pair.

This "total ionisation" of the gas is the result of the propagation of avalanches by the UV radiation caused by rearranging the deep layers of strongly ionised gas atoms. This is the "Geiger-Müller" operating regime.

A Geiger-Müller counter cannot be used to perform energy spectrometry; it is, however, very interesting and easy to implement for counting, since it delivers a high amplitude signal which does not require any additional sophisticated electronics. It does, though, have a high temporal inertia which does not allow the measurement of large amounts of radiation (see Further information below).

Further information: Geiger-Müller counters and dead time

Constitution and operation

The most common design for the measurement of surface radioactivity is the "bell"-shaped counter, consisting of a metal cylinder (cathode) in whose axis a conductive wire (anode) is mounted. This assembly is filled with an inert gas (a Helium and Argon mixture) at a pressure below the atmospheric pressure, with traces of organic or halogen vapors. The operating voltage is usually between 100 and 3000 V depending on the size of the counter.

These counters are designed for the detection of α or β particles, and the thickness of their window must be less than the range of these particles. This "window", which makes up one end of the cylinder, consists for example of a mica or Mylar sheet approximately 1.5 mg.cm^{-2} thick. The transmission of these windows depends on the radiation energy; however, in contact with a surface, it often allows to measure α and β radiation from an energy of approximately 100 keV. Probes constructed of this type of counter are called "soft beta."

We have seen that each primary ionisation produces a large series of avalanches which leaves a certain distribution of positive ions around the anode. When the potential difference between the electrodes increases, the gaseous amplification and the density of positive ions increase as well. This density creates a space charge effect as if instead of the anode, a much larger positive cylinder were present. The electric field responsible for the gaseous multiplication is altered and a saturation effect appears. In this situation, the detection pulse, if it can still be created, is of very low amplitude. The transfer of positive charges to the cathode is facilitated by the traces of organic or halogen gases, whose role is also to capture UV radiation without generating a new electron and thus a new avalanche ("quenching" effect).

Returning the gas to electric equilibrium can also be managed eventually by sharply reducing the high voltage for one or two milliseconds to prevent counter saturation due to dead time (see below).

Concepts of dead time and resolving time

We can consider the release of energy resulting from the interaction of the radiation with the detector medium as an instantaneous phenomenon. However, the information resulting from the detection chain takes longer to be delivered. In the case of a Geiger Müller counter, during the movement of the positive ions towards the cathode, the anode potential is sufficiently lowered that any radiation entering the counter does not trigger the avalanche process. The counter is thus inhibited for a period ($\sim 10^{-4}$ s) corresponding to the migration time of the ions' positive charges to the sheath. The expression "**dead time**" is used to characterise the detector medium inertia.

However, another parameter called **resolving time** τ is generally preferred since it is characteristic of the entire detection chain, not only the detector medium. It is defined as the interval between two interactions necessary to record these as two distinct events.

The existence of resolving time results in counting losses and excludes a reliable detection at a high counting rate.

e) **Region V**: the counter becomes unstable: electronic breakdowns occur and permanently damage the counter.

Because of the many technological specifications necessary to achieve efficient operating in each regime, it is not possible to manufacture a counter that can operate equally well in the three regimes: ionisation chamber, proportional and Geiger-Müller.

5.1.3. Semiconductor detectors

These detectors are the product of more recent developments than the ones described above. Their properties, in particular their high energy resolution, have allowed them to supplant scintillation and gas-filled detectors for many laboratory applications, especially in the area of alpha and gamma radiation spectrometry.

The physical phenomenon exploited in this type of detectors is ionisation.

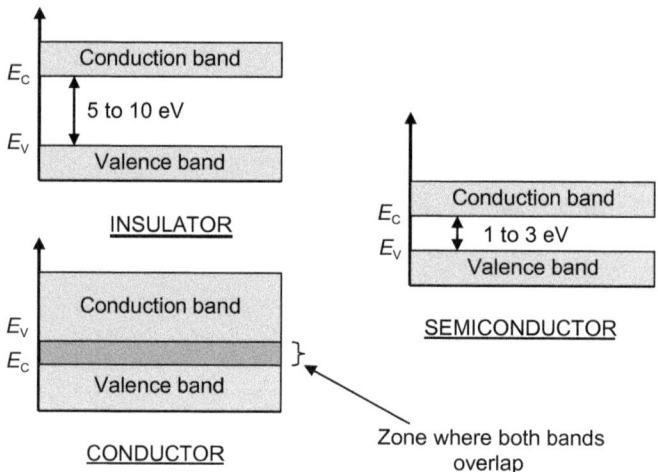

Figure 5.6. Energy levels of the least bound electrons in the different types of crystalline material (insulator, conductor, semiconductor).

In a perfect crystal structure, defined by a pattern composed of one or more atoms, the electrons have binding energies with values in well-defined ranges, called valence **bands** and demarcated by **forbidden bands** (or band gaps). The last band is called the conduction band ("free" electrons). Bands indicate the dispersion of the peripheral electrons' binding energies due to their being shared between the atoms constituting the crystal mesh.

Electrical conductivity results from the presence of electrons in the conduction band. Depending on the width of the last forbidden band, the product will be an insulator, a semiconductor or a conductor as shown in Figure 5.6.

For a semiconductor at a temperature of zero Kelvin, the valence band is full and the conduction band is empty; it behaves as an insulator. As the temperature increases, a growing number of electrons can, because of thermal agitation, spontaneously jump from the valence band to the conduction band; the material then behaves as a conductor.

On a theoretical level, the operating principle of this type of detector is simple. Each ionisation created by the radiation releases an electron (which then "jumps" in the conduction band) and a positive ion (called a "hole"). As a result of an electric field generated by a potential difference applied between two sides of the semiconductor, electrons are collected, while the corresponding holes gradually migrate, creating an electric pulse. However, this is only possible when the semiconductor material is made in the form of a backward-biased diode. It blocks the current of the voltage generator connected to its terminals, but as a result of the generator voltage, it allows the flow of the charges generated by radiation interaction induced ionisation (see Further information below).

Further information: semiconductor detectors

The semiconductor materials most commonly used in radiation protection are silicon (Si), germanium (Ge) and cadmium telluride with or without associated zinc (CdTe or CdZnTe).

No semiconductor material is totally pure. Depending on the nature of the main impurity, semiconductors are of *n*-type (electron donor impurities, such as phosphorus, arsenic, respectively for Si and Ge) or *p*-type (electron greedy impurities, for example: aluminum, gallium, respectively for Si and Ge). These semiconductors are known as extrinsic. The impurities' presence increases, and even enables the conductivity of the material, which is called "extrinsic" when impurities are intentionally added.

Two semiconductors, one *p*-type and the other *n*-type, appended to each other constitute what is called a junction, or diode, in which a region empty of charge carriers appears. Connecting the two semiconductors, initially with each being electrically neutral, results in the diffusion of the majority electrons from the *n* region to the *p* region, and conversely, the diffusion of the majority of "holes" from the *p* region to the *n* region. In the contact zone, holes and electrons only appear when they are created by an external cause, for example the interaction of an ionising radiation. Everything occurs as if in the presence of a "**solid ionisation chamber**."

Compared to a gas ionisation chamber, the semiconductor detector offers two important advantages:

- for equal volumes, a detection efficiency much larger than that of a gas ionisation chamber (the density being about 2000 times the density of a gas at atmospheric pressure);
- the average energy required to create an electron-hole pair is about 3 eV; for an equal amount of energy transferred by radiation, the semiconductor provides approximately 10 times more ion pairs than a gas ionisation chamber, which leads to a much smaller relative statistical fluctuation in the resulting electrical signal and allows for a very good energy resolution with spectrometers.

However, two drawbacks are worth noting:

- due to the low energy of the forbidden band, the use of germanium detectors for gamma spectrometry requires cooling them, which calls for the presence of a liquid nitrogen tank or the use of a "cryogenerator";
- it remains difficult to obtain large sized detectors at reasonable prices.

Semiconductors are primarily used in spectroscopy when it is necessary to work with a good energy resolution. They are also used in operational dosimetry.

All of the types of detectors that were presented so far, when combined with appropriate electronic units, can structurally:

- count the incident radiation in "real time";
- perform (for some of them) spectrometry;
- in general, instantly measure the absorbed dose rate.

This is not the case for detectors that are discussed later in this first part of the chapter.

5.1.4. Photographic emulsions

This is the simplest and oldest "detector". In 1896, Henri Becquerel discovered natural radioactivity by observing the blackening of photographic plates placed in the vicinity of uranium salt, soon after Wilhelm Conrad Röntgen performed the first radiograph using the X-rays he had discovered in late 1895.

Photographic emulsion consists of crystallised grains of silver bromide (AgBr) and interstitial Ag^+ ions suspended in an organic gel ("nuclear" emulsion) or deposited as a "film" on an organic substrate.

The interaction of ionising radiation with the AgBr grains results in bromine releasing electrons which are lost to the interstitial silver and lead to its reduction $Ag^+ + e^- \rightarrow Ag$. A seed of metallic silver is thereby formed. Together, these seeds constitute the latent image. The development process can significantly increase the number of silver atoms to the point of forming metallic grains which can be observed under a microscope and induce an average blackening of the emulsion proportional to the number of radiations that have interacted with it.

In nuclear emulsions, the traces are microscopically counted and analysed to determine the number of radiations that have interacted with the gel and the significance of their linear energy transfer. In practice, these detectors can be useful for measuring fast neutrons which, through inelastic scattering, result in recoil protons whose energy deposit is "mapped" into the gel.

A film instead measures the optical density (OD) which is an indicator of the average blackening of the photographic emulsion as shown in Figure 5.7.

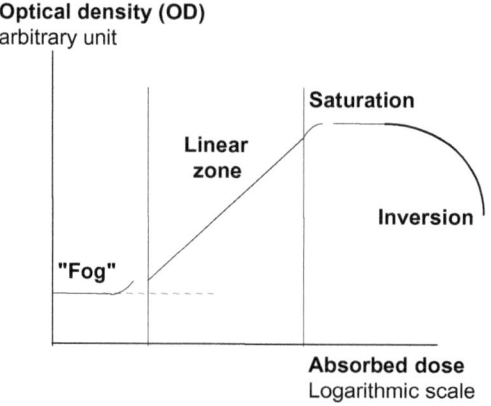

Figure 5.7. Blackening curve of a photographic emulsion.

This optical density can be linked to the absorbed dose thanks to a calibration curve, thus allowing a quantitative measurement after exposure to X-rays, γ radiation and electrons.

At low doses, the natural "fogging" of the film, due to spontaneous ionisation of the AgBr grains caused by the electrons' thermal agitation, leads to a sensitivity threshold of about 0.2 mSv. At high doses, first a saturation effect occurs, then an inversion effect; therefore the range of use is located in the "rising" section of the curve, as shown above. Using photographic emulsions of different sensitivities allows extending the range of use and the range of measurable doses.

In addition to this first drawback (saturation, inversion), we must consider the fact that its response (in degrees of OD per mGy) greatly depends on the energy of the X-rays and γ radiation and, in particular, is hypersensitive to low energies (the same absorbed dose causes 30 times more blackening at 50 keV than at 1000 keV).

To mitigate this over-response and ensure the energy-independence of the film's response as much as possible, the dosimeter film is placed behind carefully chosen filters. This "energy compensation" technique is implemented whenever a detector exhibits a response that varies too much as a function of the radiation to be measured, as discussed later.

Because they are sensitive to light, film and emulsion must be packaged in sealed envelopes.

The photographic dosimeter has long been the most widely used passive dosimeter, having, for that matter, a regulatory nature. The lowering of regulatory limits for yearly average equivalents dose and effective dose and the subsequent need for measurements of low levels of irradiation, gradually led to the disuse of dosimeter film for X-rays, γ radiation and electrons, and of nuclear emulsions for neutrons, due to their high level of intrinsic noise.

They are being replaced, in particular, by radioluminescent dosimeters, that are described below.

5.1.5. Radioluminescent detectors

As described earlier, the crystal structure is a lattice in which the least bound electrons are distributed in the valence band and can be released into the conduction band when radiation induced ionisation occurs.

During the release of energy by radiation in a crystal, the large number of ionisations generate a large number of electron-hole pairs. In the absence of an electric field to cause them to drift, or if the crystal is not conductive, the electrons and holes tend to recombine quickly.

Radioluminescent detectors are materials in which impurities, whose nature and dosage are carefully chosen, are at the origin of intermediate energy levels in the crystal's forbidden band. These intermediate levels permit a stable attachment of electrons and holes: they are called "traps". These traps can capture a fraction of the electrons and holes generated by radiation interaction induced ionisation, thus preventing them from recombining spontaneously in the valence band.

The energy levels of these traps are sometimes such important that trapped electrons and holes cannot leave their trap spontaneously, as opposed to what happens in scintillators.

Therefore, when radiation interacts, some of the products of ionisation becomes trapped. To "release" electrons and holes from their traps, it is necessary to provide external energy:

– heat in the case of thermo-luminescence;

– light in the case of the photo-luminescence and in the case of optically stimulated luminescence.

The result of energetic stimulation of the irradiated material is the emission of light which corresponds to:

- the recombination of trapped electrons and holes, in the case of thermo-luminescence and in the case of optically stimulated luminescence;
- the de-excitation of electrons which are excited in their trap by ultraviolet light, in the case of photo-luminescence.

The amount of light emitted is measured by a photomultiplier whose average current is read.

After calibration, the amount of light collected is related to the absorbed dose.

5.1.5.1. Radio-thermo-luminescent detectors

In the radio-thermo-luminescent detectors, traps are usually stable at room temperature, however the electrons can return to their original state through simple heating (between 110 °C and 260 °C depending on the material). This heating process is accompanied by the emission of light photons, the number of which varies with the amount of energy transferred to the crystal by the radiation shown in Figure 5.8 below.

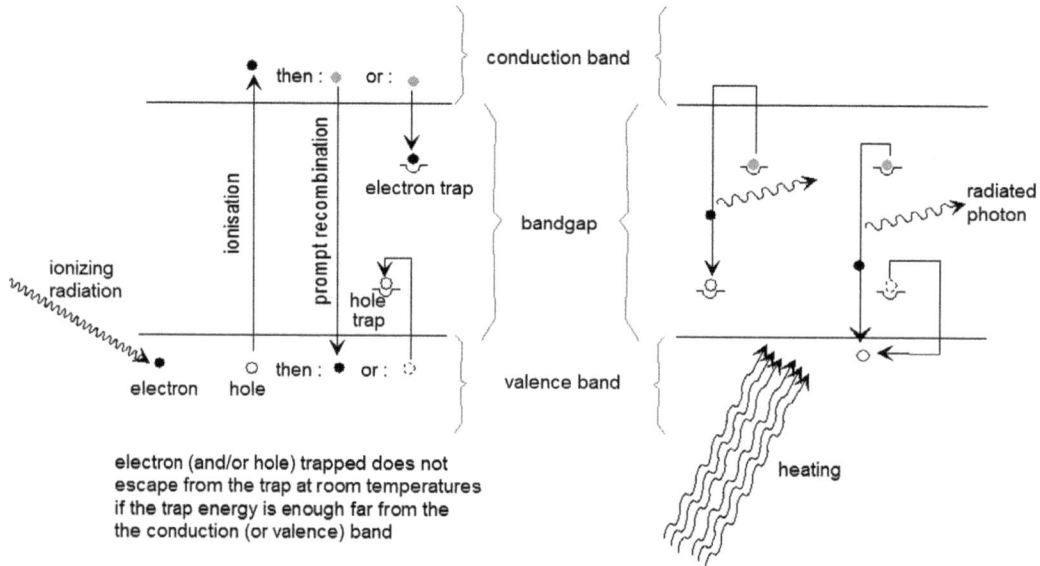

Figure 5.8. Diagram of the operating principle of radio-thermo-luminescent detectors.

These detector materials, such as lithium fluoride, FLi, doped with (among others) magnesium, are used to measure the absorbed dose in a very wide range (50 µGy to 2000 Gy). They are considered equivalent to living tissues with regards to electromagnetic radiation because the average atomic numbers are comparable. Because of their small size, they can also make excellent "contact" measurements, and are very useful for performing extremities dosimetry.

The use of radio-thermo-luminescent detectors (TLD) containing lithium fluoride, LiF, doped with impurities of Ti, Mg, Cu, etc. responsible for the traps, is widespread. This material is usually preferred over others because its average atomic number is close to that of human tissue and the traps are fairly stable over time (the "fading" is weak), which allows integration periods of several months without significant loss of data.

LiF dosimeters enriched with ^{6}Li can be used to measure thermal neutrons, for which this isotope has a high absorption cross-section.

The nuclear reaction ^{6}Li$(n, \alpha)^3$H allows the use of the internal converter that is ^{6}Li by integrating the ionisation due to the kinetic energy of the reaction products, i.e. the α particle and the ^{3}H nucleus.

Because ^{6}Li is inherently present in natural lithium, it is necessary to use LiF detectors enriched with ^{7}Li, which is insensitive to neutrons, in order to make a differentiated measurement of doses due to photons from those due to neutrons in mixed irradiation fields.

Dosimeters made from FLi overestimate doses for photon energies of a few dozen keV; therefore they are utilised as badges composed of several small pellets arranged behind a set of filters intended to compensate for this over-response, in a manner similar to old dosimeter films.

Detectors being subject to extreme heating necessitates the use of powders in the form of sintered pellets or pellets embedded in a PTFE matrix.

5.1.5.2. Radio-photo-luminescent (RPL) detectors

The fluorescence radiation emitted by certain substances in ultraviolet light exhibits well defined characteristics which are modified after irradiation. Calibration allows the linking of fluorescence of the irradiated material with the absorbed dose received.

These detectors are generally over-sensitive for low energies (often by a factor of 2 to 3). This disadvantage can be compensated for by using appropriate filters, using the same technique than dosimeter films.

The new passive dosimeter is based on this radio-photo-luminescence (RPL) technique and uses a fluorescent glass detector doped with silver. Its operation is described in Figure 5.9.

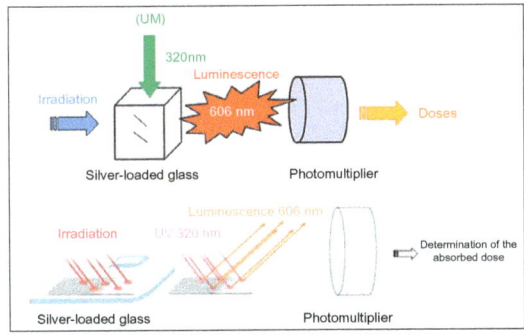
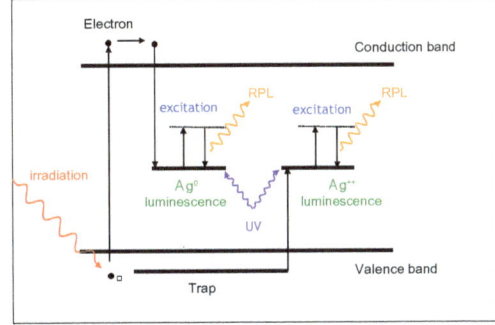

Figure 5.9. Diagrams of the operating principle of radio-photo-luminescent detectors (sources IRSN and CEA).

This dosimeter has an extremely low detection threshold (a few dozen µGy). This is a significant step forward compared to dosimeter film. It also offers excellent angular response and energy response through compensation filters.

The RPL technique allows the reading of the dosimeter as often as needed without any loss of information.

In addition, by scanning the fluorescent glass plate's surface with the ultraviolet source, it is possible to obtain an "image of the dose" and to thus refine information on the radiation.

Dosimeters can be fully reset by high temperature heating.

5.1.5.3. Optically stimulated luminescence (OSL) detectors

This technique relies on the same principle as radio-thermo-luminescence.

It differs only in the method of luminescence stimulation. Instead of heating the detector, it is exposed to a very intense (and often pulsed) light source such as a laser, giving the name to the technique: "Optically Stimulated Luminescence."

Luminescence is also measured by a photomultiplier equipped with a filter to not measure the stimulation light.

Carbon-loaded alumina, Al_2O_3, is widely used and allows the measurement of single absorbed doses of a few µGy.

Unlike radio-thermo-luminescence, reading the detector does not entirely empty all levels of trapping: a second reading is still possible.

Detectors can be reset by a long exposure to an intense light source.

The fact that this method does not need high temperatures in the detectors allows the use of polymer matrices.

5.1.6. Other types of detectors

Beside the most frequently used detectors described above, other detectors with more specific uses can also be used and are described thereafter.

5.1.6.1. Solid state nuclear track detectors

Radiation with high linear energy transfer, such as alpha radiation, protons and other recoil nuclei, cause significant disturbances along their path in all materials because of their very high density of interactions. When using some organic films, the tracks left by each interaction can be plainly visible under an optical microscope at a very high magnification. The size of these tracks can be amplified by chemical or electro-chemical etching of the film.

After exposure, the number of tracks is proportional to the radioactivity concentration, or to the dose equivalent after calibration.

Tracks are counted with an optical microscope equipped with a camera whose images are processed by a dedicated analysis and processing software.

The most commonly used materials are coloured cellulose nitrate (trade name: LR115) and allyl diglycol polycarbonate (trade name: CR 39).

They allow the measurement of α emitting radionuclides in their vicinity; therefore, they can measure solid decay products of atmospheric radon, as well as fast or thermal neutrons, depending on the nature of the converter layer(s) "glued" to the detector film.

They are widely used for the dosimetry of radon, to monitor internal dosimetry in uranium and thorium mines.

They are most frequently found as complements to RTL, RPL or OSL dosimeters to evaluate the dose equivalents due to neutrons.

5.1.6.2. Activation detectors

Among ionising radiations commonly encountered in the practice of radiation protection, excepting high energy γ radiation acting on some small nuclei, only neutrons are capable of producing nuclear reactions. Among the nuclei transformed by these reactions, only some become radioactive. Some materials exhibit a neutron capture property which is used to measure high neutron fluxes in the context of accidents, particularly during criticality accidents. Workers are equipped with belts, and the premises with pellets that include a set of materials which can be activated depending on the energy of neutrons with various cross sections.

In case of a neutron irradiation accident, the objects worn by the victims are used to estimate the dose equivalent in the accident reconstruction. This process often resorts to analysis of gold and silver jewelry, metallic prostheses, etc.

5.1.6.3. Bubble detectors

These detectors are dedicated to the measurement of neutrons, and belong to the family of phase-change detectors, which have been used very early on in radiation measurements to follow their tracks (in particular the Wilson chamber). The principle relies on the fact that by interacting, the radiation deposits an amount of energy that is very small, yet large enough to turn a saturated vapour into liquid, or a liquid into gas along the path of a directly ionising radiation.

Bubble detectors take the form of a transparent tube containing an organic gel, also transparent, which itself contains several tens of thousands of microscopic freon droplets near the point of evaporation. When a neutron interacts in the gel, it hits a proton which recoils. If this proton encounters a droplet, the energy it carries causes the droplet to evaporate: the droplet then becomes a gas bubble in the gel. The number of bubbles thus generated is proportional to the number of neutrons having interacted, or after calibration, to a dose equivalent.

The detector can be reset by pressurising the gel, which is accomplished by twisting the cap at the end of the tube.

This type of dosimeter, being especially sensitive to temperature, is less frequently used in favour of solid state nuclear track detectors (see section 5.1.6.1).

5.1.6.4. Chemical detectors

In these detectors, ions and free radicals produced by radiation can chemically react to form new compounds, the amount of which being related to the energy absorbed.

Such detectors are not very sensitive, and therefore usable only for high absorbed doses, for example in intense radiation beams. They are rarely used in radiation protection.

5.2. Electronics associated with detectors

The signals delivered by detectors must be subjected to specific electronic processing.

In scintillators, proportional and Geiger-Müller counters or semiconductor detectors, the radiation detection is ultimately reflected in the occurrence of an electron "burst", called electrical pulse. Electrical detection pulses, each corresponding to one radiation interaction, must be correctly tallied for the purpose of counting. In the case of energy spectrometry, which provides more information than simple counting, the amplitude of each pulse must also be analysed in order to determine to which energy in the measured spectrum the detected "count" should be allocated. The number of counts measured in a region of the spectrum is proportional to the number of radiations having interacted by releasing the corresponding energy.

In ionisation chambers (excepting grid and fission chambers), the principle is to measure an electric current, usually of low intensity, which reflects the interactions of a radiation flux.

The detectors' luminescence after irradiation is measured by a photomultiplier whose output current is measured.

In the case of very low currents, electrometers are used.

The intensity of the average electric current is proportional to the radiation flux, integrated or not over time.

Other types of detectors provide images that need to be analysed in terms of tracks, bubbles, etc. Their number is proportional to the radiation flux integrated over the duration of the exposure.

In the case of detection pulse processing, one of the peculiarities of "nuclear electronics" is that each individual pulse must be accurately measured. Indeed, the process that produced the radiation responsible for the interactions in the detector is a completely random process in which a second radiation has a very high probability of being emitted immediately after the first radiation (Poisson process).

Moreover, the detection pulses sometimes have very low amplitudes, especially with very low energy radiation. It is therefore necessary to clearly distinguish them from the electronic noise generated by the electrons' thermal agitation in circuits and components. This electronic background of random amplitude contributes to the fluctuations in the measurement of energy in spectrometry and thus degrades the spectrometer resolution. It is therefore important to filter it in frequency domain in order to avoid having too many pulses on the 0V base line (LF noise), or the noise combining with the pulses themselves (HF noise).

It is also important to clearly distinguish this noise in amplitude in order to avoid counting events that do not correspond to radiation interactions.

Depending on the information sought and which detector is used, a measurement chain's sophistication is variable; it includes several modules among those shown in Figure 5.10.

5 – *Detection and measurement of ionising radiation*

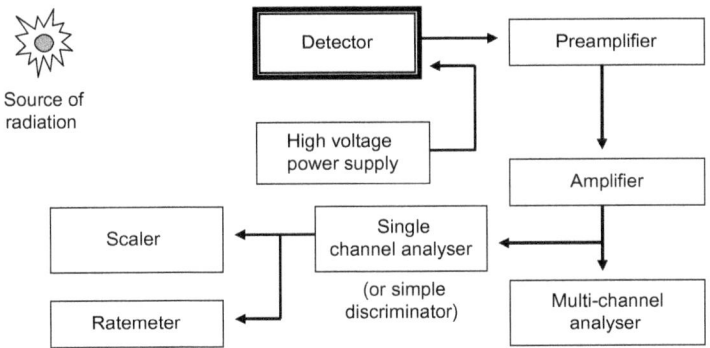

Figure 5.10. Nuclear electronics modules used to compose a measurement chain.

First, a **high voltage power supply** allows the delivery of a constant electric potential which is applied to the detector or the photomultiplier. It is used to collect or multiply the electrons generated on the detector's side.

One or more low voltage power supplies, not shown in the figure, are necessary to support the modules.

A **preamplifier** is generally placed directly after the detector. It can slightly amplifying pulses proportionally, if these are very weak. It is used to form a first pulse which can be transmitted in the cables to the next module. In addition, its components and circuits are very precisely designed to get the best "detection signal / electronic noise" ratio. A typical post-preamplifier pulse is shown in Figure 5.11.

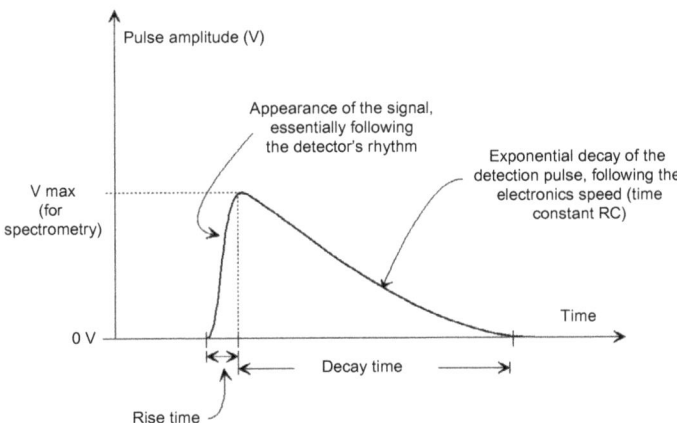

Figure 5.11. Preamplifier output detection pulse.

An **amplifier** linearly increases the amplitude of the pulses obtained from the preamplifier. The amplification factor (gain) can be adjusted so that the output pulses remain within the operational range of the electronic units further down the line (often 0 to 10 V). When the amplifier is used to perform pulse amplitude spectrometry, it also plays a sophisticated role in shaping pulses, which allows the filtering of electronic noise and produces a pulse peak that can be accurately measured.

A **discriminator** is usually present in the chain; it allows retention in the detection event, counting only pulses whose amplitudes are greater than an adjustable voltage, known as the threshold. Its role is mainly to eliminate electronic noise peaks that could be confused with very small detection pulses. In some proportional counters, it is also used to separate pulses caused by the detection of gamma radiation from neutron-induced pulses.

A **single-channel analyser** selects pulses whose amplitude lies between two voltage thresholds. It is composed of two discriminators whose threshold settings allow the definition of a voltage window. If the V_{max} amplitude of a pulse is within the window, it generates a "logical" pulse as a rectangular, "0 – 5 V – 0" wave lasting about 500 ns. These logical signals are to be counted by a "scaler", or their number is to be averaged over time by a "ratemeter".

The **multi-channel analyser** is equivalent to n single-channel analysers with the same window width, connected in parallel. It allows recording of a spectrum of pulse amplitudes in a single acquisition. Its principle relies on the conversion of the maximum amplitude of each pulse into a number comprised between 1 and 2^n where n is the number of bits of the voltage converter (ADC). Unlike in the digitisation of periodic signals, such as music, here we do not digitise the entire shape of the signals, but only their maximum amplitude. The converter is associated with a memory of 2^n segments called "channels". The number resulting from the conversion designates the number of the memory channel where the detection count will be incremented.

Nowadays, multichannel analysers are often found as equipment connected to a computer (through a USB connection, for example). This computer runs specialised software that allows the spectrum acquisition, the display of spectra by retrieving the contents of the analyser's memory channels, and various calculations required to make use of the spectra.

The **scaler** is an electronic device that allows either the counting of logical pulses for a given time (pre-time mode) measured by an internal electronic timer, or the indication of the time that must elapse in order to obtain a predetermined number of pulses (pre-count mode).

In certain circumstances, what we are interested in is the direct reading of the average count rate. It is then necessary to use an electronic module designed for such measurements, called a "**ratemeter**".

A ratemeter can mitigate the effect of statistical fluctuations observed in the counts by implementing large time constants τ_{icto} in order to average low pulse fluxes.

Figure 5.12 shows an example of a portable ratemeter which also incorporates a high voltage power supply and an amplitude discriminator (these functions can also be implemented in the probe detectors connected to the device).

Combining the electronic units described above allows the setting up of detection chains. Not all of the components are strictly necessary. For example, the single-channel analyser is not required to perform countings.

In the specific case when the detector is a Geiger Müller counter, because of the high amplitude of the delivered pulses, the preamplifier is a simple impedance adapter. The amplifier and single-channel analyser are unnecessary, provided that a simple pulse shaping and calibration circuit is available. Such a chain is therefore inexpensive, and yet remains competitive in fields such as the detection of energetic electrons, even at low levels.

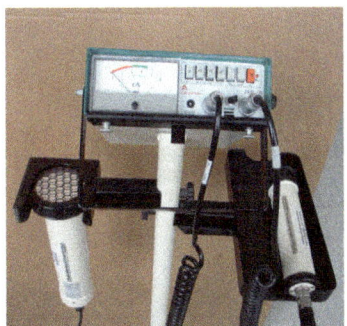 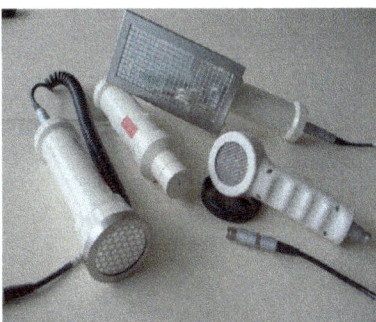 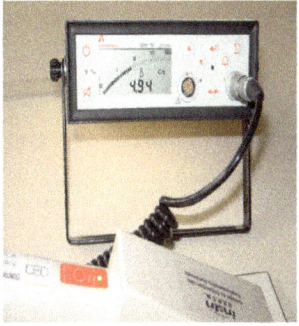

Figure 5.12. left: Mini portable ratemeter (MIP) manufactured by Canberra, photographed here with a "soft beta" type probe detector. **Center right:** alpha probes, soft beta, alpha, beta and gamma generation. **Right:** MIP Digital (Sources INSTN).

5.3. Measurement methods and practices

Measurement may involve pulse counting or current measurements of varying sophistication.

5.3.1. Detection pulse counting

5.3.1.1. General considerations: counting and counting rate

As discussed above, when radiation interacts in an active detector capable of delivering a different signal for each interaction, an "electron burst" occurs and is transformed into an electrical detection pulse. This pulse is commonly called "**count**".

The number N of counts tallied during a period of t seconds is called a "**counting**".

When the counting is performed without specifying the pulse amplitude, the term used is "**total counting**". Otherwise the counting is performed in a particular region of interest in a spectrum, such as a peak (which corresponds for example to the scenario of total absorption of the energy of radiation emitted as an energy line).

The ratio of counts over time is called the "**counting rate**", written $\dot{N}$ or n.

It is expressed in "counts per second" which is written as: s^{-1}, c/s, pps or even cps.

$$\dot{N} \text{ or } n[s^{-1}] = \frac{N}{t[s]}$$

The counting of N radioactive events is subject to statistical fluctuations, modelled by the Poisson distribution. This distribution, whose variance is equal to the mean, allows the approximation of the standard uncertainty $u(N)$ of a counting N by the square root of its value, $\sqrt{N}$:

$$u(N) = \sqrt{N}$$

From this, we can infer that for a counting N, the relative fluctuations vary as: $\sqrt{N}/N$.

Therefore, countings which yield 4 counts on average fluctuate by 50% from one counting to the next.

A counting that yields 100 counts fluctuates by approximately 10% from one measurement to the other.

A counting of 10 000 counts corresponds to relative fluctuations of about only 1%.

This estimate of the uncertainty $u(N) = \sqrt{N}$ applies only to countings, as the uncertainty must have the same dimension as the observed value, and the Poisson distribution only models dimensionless physical values.

For example, for the counting rate n: $u(n) \neq \sqrt{n}$; $\quad u(n) = \frac{\sqrt{N}}{t} = \sqrt{\frac{n}{t}}$

Note: When the counting rate is obtained through a ratemeter whose time constant is τ_{icto}, the uncertainty $u(n)$ of the counting rate n is $u(n) = \sqrt{\frac{n}{2 \times \tau_{icto}}}$

The measurement of the counting rate can allow the detection of the presence of radionuclides in background radiation if the detector is adapted to the interactions of the radionuclides' radiation. It is even possible to measure the activity of a given radionuclide.

Measuring a counting rate may also give information about a quantity such as an absorbed dose rate once the response of the detection chain has been characterised with a standard radiation beam.

To switch from a counts per second measurement either to the activity of a radionuclide (responsible, for example, for surface contamination) or to a dose equivalent rate, the following three corrections should be made:

– counting losses due to dead time;

– background radiation and background effect of the detection assembly;

– measurement efficiency.

5.3.1.2. Correcting for counting losses due to dead time

As described above in the note on the Geiger-Müller counter, all detection systems have a temporal inertia.

This temporal inertia is due either to the detector, which takes some time to return to its initial state after an interaction (e.g. Geiger counter), or to the electronics, which take some time to produce the count or to classify it within a multi-channel spectrum.

This inertia causes a global resolving time τ, which is equal to the minimum time interval that must separate two detected events in order to produce two distinct pulses: below τ only the first event is recorded.

For simplicity's sake, in that document, the term "dead time" is used instead of resolving time.

There are two simple models of the complex phenomenon of the dead time.

The first model, which actually concerns detectors, considers that any new detection event extends by τ the period during which the detector generates a pulse. During this period τ, it is unable to generate a second pulse, while remaining sensitive to the following interactions that can still extend the detector's unproductive period by τ. Geiger-Müller counters do not function according to this model. In this model, when the measured counting rate n_c is represented as a function of the interaction rate n_i, the result is a relationship that goes through a maximum and then tends to decrease to zero. Such detectors are not well suited for the measurement of high levels of radiations because when the interaction rate is very high, they can produce a response equal to zero. One solution to counter this type of flaw is to sharply reduce the high voltage for approximately two milliseconds to avoid saturation of the detector: this essentially makes the dead time a fixed length of time, which is then easier to correct quantitatively.

The second dead time model concerns the electronics associated with the detector. The dead time is assumed to be fixed in nature, i.e. if a first interaction generates a period τ necessary to produce a count, during τ no new interaction event is taken into account. It has no influence on the period τ during which the chain cannot produce a second pulse, because the chain is then occupied producing the first.

In this model of "fixed dead time" the rate of interactions n_i is related to the counting rate n_c by the equation:

$$n_i = \frac{n_c}{1 - n_c \tau}$$

5.3.1.3. Correction of the background from the detection assembly

In the absence of a radiation source deliberately placed in front of a detector, we notice that the detector produces a counting rate which varies depending on the location and chosen detector.

This counting rate, called "**background**", derives from two sources.

The first cause is the "background radiation", which is the flux of ionising radiation surrounding the detector. There are multiple origins of these radiations:

- radiation emitted by natural radionuclides present in soil and in building materials;

- radiation emitted by radionuclides that are naturally present in the atmosphere – either as descendants of the isotopes of radon, a radioactive gas that emanates from soil and building materials – or as by-products of the interaction of air with cosmic radiation;

- cosmic radiation;

- radiation resulting from human activities.

These various causes explain why the background radiation varies according to the location, and consists mainly of gamma radiation.

The second source of noise is inherent to the detection chain, independant of any external radiation.

Pulses can arise spontaneously in the chain in the absence of external radiation interacting with the detector.

This phenomenon, called "background effect", may be caused by:

- radioactive impurities contained in the materials composing the detection system;

- spontaneous electron removal due to thermal agitation in the vicinity of emissive surfaces, or in the photomultipliers where a "dark current" is systematically observed;

- the poor differentiation of some peaks of electronic background.

"Background radiation" and "background effect" can thus randomly combine to form the so-called "background".

This background is included in any radiation measurements, making them "raw" n_i measurements, even after adjusting for the effects of dead time.

We should therefore subtract the background n_b to obtain the net counting rate n_{inet}:

$$n_{inet} = n_i - n_b$$

Note: The background n_b is generally low enough to neglect losses due to the associated dead time.

5.3.1.4. Uncertainty on the net counting rate and decision threshold

Statistical fluctuations of the net counting n_{inet} result both from the fluctuations of the raw counting adjusted for dead time, n_i, and from the statistical fluctuations of the background n_b.

Both fluctuations can be modelled by a Poisson distribution, so that the standard uncertainties associated with each countings are:

$$u(n_i) = \sqrt{\frac{n_i}{t_i}} \text{ and } u(n_b) = \sqrt{\frac{n_b}{t_b}}$$

where t_i and t_b, respectively, are the raw counting time and background counting time.
Note: we do not usually take into account the uncertainty resulting from the dead time induced correction to calculate $u(n_i)$.

The combination of these two standard uncertainties $u(n_i)$ and $u(n_b)$ yields the combined standard uncertainty $u_c(n_{inet})$. n_{inet} resulting from a difference $u(n_{inet})$ is calculated by:

$$u_c(n_{inet}) = \sqrt{u^2(n_i) + u^2(n_b)} = \sqrt{\frac{n_i}{t_i} + \frac{n_b}{t_b}} = \sqrt{\frac{n_i + n_b}{t}} \quad \text{if } t = t_i = t_b$$

With a ratemeter whose time constant τ_{icto} varies according to the range of counting rate measured, we have then:

$$u_c(n_{inet}) = \sqrt{\frac{n_i}{2\tau_{icto-i}} + \frac{n_b}{2\tau_{icto-b}}} = \sqrt{\frac{n_i + n_b}{2\tau_{icto}}} \quad \text{if } \tau_{icto} = \tau_{icto-i} = \tau_{icto-b}$$

When there is no radiation to be measured besides the background radiation, then: $n_i = n_b$ and $n_{inet} = 0$.

However $u(n_{inet}) \neq 0$; indeed: $u(n_{inet}) = u(0) = \sqrt{\frac{n_b + n_b}{t}} = \sqrt{2n_b/t}$

This standard uncertainty reflects fluctuations in the background that can hide small sources of radiation. It is used to help determine whether or not an observed counting rate exceeds the background.

While a raw counting follows a Poisson statistical distribution, a net counting rate such as n_{inet} follows a Gaussian distribution. Accordingly, 95% of the values are located at the centre of the curve, 2 standard deviations above and below the mean value. In other words, 2.5% of the values fall below "$n_{inet} - 2u(n_{inet})$" and 2.5% of the values fall above "$n_{inet} + 2u(n_{inet})$".

This allows the calculation of a "decision threshold" which is associated with a risk α, called "false alarm" or "false positive" of 2.5% at the value:

$$DT_{\alpha=2.5\%} = 0 + 2\sqrt{2n_b/t} \approx 2{,}83\sqrt{n_b/t}$$

With a 5% chance of incorrectly concluding that the counting rate is significant, i.e. if we end up observing only unlikely fluctuations of the background:

$$DT_{\alpha=5\%} = 1{,}645\sqrt{2n_b/t} \approx 2{,}33\sqrt{n_b/t}$$

With a ratemeter of time constant τ_{icto}, we have: $DT_{\alpha=5\%} = 1{,}645\sqrt{2n_b/\tau_{icto}}$

When $n_{inet} > DT_\alpha$, we can conclude that n_{inet} is significant to the risk α, and that its value can be obtained by estimating its standard uncertainty $u(n_{inet})$, which increases as n_{inet} approaches DT_α.

When $n_{inet} < DT_\alpha$, we can conclude that n_{inet} is not significant. Here, there remains a risk β, called risk of "non-detection" or "false negative", that the net counting rate observed is caused by radiation inducing, on average, a counting rate $\bar{n}_{DL}$. This average counting rate, which is expressed with the low probability β below DT_α, is at the origin of the detection limit which is discussed in section 5.3.1.9.

5.3.1.5. Application of a measurement efficiency to determine the activity of a radionuclide

In cases where the counting rate n_{inet} (adjusted for dead time and background) is caused by a particular radionuclide, n_{inet} is proportional to the activity A of this radionuclide.

The proportionality factor is called "activity efficiency" and written ε_A:

$$\varepsilon_A = \frac{n_{inet}}{A} \quad \varepsilon_A \text{ is expressed in (c/s)/Bq or } s^{-1}.Bq^{-1}$$

ε_A is determined by calibration, by exposing the detector to the radiation flux of a source with a calibrated activity and containing only the specific radionuclide.

ε_A is therefore a "calibration factor" valid only for the radionuclide in question, the detector being used, and in measurement conditions similar to those of the calibration.

Once ε_A is known, it is used to determine the radionuclide's activity in measurement conditions closest to those used to determine ε_A:

$$A_{inc}[Bq] = \frac{n_{inet}[c/s]}{\varepsilon_A[(c/s)/Bq]} \quad \text{with} \quad \frac{u_c(A_{inc})}{A_{inc}} = \sqrt{\left(\frac{u(n_{inet})}{n_{inet}}\right)^2 + \left(\frac{u(\varepsilon_A)}{\varepsilon_A}\right)^2} \quad \text{(see below)}$$

For example, if ε_A is determined for a given detector, 3 mm away from a cobalt-60 source with a surface of radioactive deposit of 1 cm² and whose radioactivity is confined between 2 mg.cm^{-2} thick plastic sheets, a valid use of ε_A requires that the source of ^{60}Co of unknown activity must have the same characteristics as the standard source, the detector must be identical and the measurement must be performed in the same conditions: actually, this is rarely verified. In practice, we can only approximate these conditions as far as possible.

When the measurement conditions differ significantly from the calibration conditions, it should be considered that the situation does not allow any measurement, but rather an assessment or rough screening.

When the radionuclide is unknown, activity is measured as **"equivalent activity"** by taking the radionuclide most likely to be present on the measurement site as the reference radionuclide. For example, activities measured on an electronuclear production site are measured in "$Bq_{eq.}{}^{60}Co$", and the efficiency is applied to ^{60}Co, even if this radionuclide is not the one being measured.

For a decay to be "tracked" by a detection "count", the following four conditions must be achieved:

- the decay must generate radiation;
- this radiation must be emitted towards the detector;
- in order to escape from the source, the radiation must not interact significantly in the source;
- this radiation must interact in the detector.

Thus, the efficiency ε_A depends on:

1. the radiation emitted by the radionuclide (nature, energy and intensity of the emission);
2. the fraction of radiation emitted towards the detector, thus the geometry of the measurement (geometric efficiency);
3. interactions of this radiation in the source containing the radionuclide;
4. interactions of this radiation in the detector.

Because radioactivity is often measured by contact with a radioactive deposit on a substrate that can only be accurately measured on one side, we use an efficiency $\varepsilon_{2\pi}$ called "interaction efficiency in 2π steradian" (in the half-space facing the detector).

This "contact efficiency" combines items 2, 3 and 4 above, and must include assumptions about the source's emission efficiency ε_S (compromise between radiation backscatter on the substrate and mitigation in the deposit). In this respect, the ISO 7503 standard suggests the use of $\varepsilon_S \approx 0.5$ for β radiation whose $E_{\beta max} > 150$ keV and $\varepsilon_S \approx 0.25$ for α and β radiation whose $E_{\beta max} \leq 150$ keV.

Again, this efficiency is determined by calibration, by exposing the detector to the radiation flux $\Phi_{2\pi}$ of a source calibrated in flux.

$$\varepsilon_{2\pi} = \frac{n_{inet}[c/s]}{\Phi_{2\pi}[rays/s \text{ under } 2\pi sr]} \; ; \text{ it is expressed in "counts per emitted radiation in } 2\pi \text{ sr"}$$

$\varepsilon_{2\pi}$ depends only on the nature and the energy of the emitted radiation.

It is used to measure unknown activities as follows:

$$A_{inc}[Bq] = \frac{n_{inet}[c/s]}{\varepsilon_{2\pi}\varepsilon_s I_{ray}}$$

where:

n_{inet} is the counting rate, adjusted for dead time and background, in "counts per second";

$\varepsilon_{2\pi}$ is the interaction efficiency in the half-space facing the detector, in "counts per radiation emitted by the radioactive deposit in 2π sr";

ε_S is the emission efficiency of the source "radiation emitted by the radioactive deposit within 2π sr per radiation emitted in the deposit in 4π sr " ($\varepsilon_S \leq 0.5$);

I_{ray} is the emission intensity of the radiation emitted by the radionuclide, in "radiation emitted by decay (in 4π sr) within the radioactive deposit".

In this case, the standard uncertainty $u(A_{inc})$, for a formula containing only products and quotients, is estimated as follows:

$$\frac{u_c(A_{inc})}{A_{inc}} = \sqrt{\left(\frac{u(n_{inet})}{n_{inet}}\right)^2 + \left(\frac{u(\varepsilon_{2\pi})}{\varepsilon_{2\pi}}\right)^2 + \left(\frac{u(\varepsilon_S)}{\varepsilon_S}\right)^2 + \left(\frac{u(I_{ray})}{I_{ray}}\right)^2}$$

Emission intensities of a given radiation are generally known with great accuracy: $u(I_{ray})/I_{ray}$ 1%. In addition, ε_S results from an assumption inferred, for example, from a standard, for which no uncertainty can be quantified.

Therefore, the calculation is generally limited to the following:

$$\frac{u_c(A_{inc})}{A_{inc}} = \sqrt{\left(\frac{u(n_{inet})}{n_{inet}}\right)^2 + \left(\frac{u(\varepsilon_{2\pi})}{\varepsilon_{2\pi}}\right)^2}$$

5.3.1.6. *Measurement of surface, volume and specific activity*

Activity calculated from measurements of the adjusted net counting rate and the estimated measurement efficiency allow the determination of an average surface activity.

This measurement can be performed directly on a flat surface if it is not exposed to excessive local background radiation originating from, for example, the radioactivity present in a parcel whose external sides are to be measured for potential radioactivity.

A thorough measurement requires that the contaminated area is larger than the detection area S_{det}, and that the detector used is not too sensitive to photon radiation emitted by the radionuclide(s) to measure, since these photons can also come from outside of S_{det}.

In this case, the **"surface activity"** is calculated as:

$$A_S[Bq/m^2] = \frac{n_{inet}[c/s]}{\varepsilon_{2\pi}\varepsilon_s I_{ray} S_{det}[m^2]}$$

with $\dfrac{u_c(A_S)}{A_S} = \sqrt{\left(\dfrac{u(n_{inet})}{n_{inet}}\right)^2 + \left(\dfrac{u(\varepsilon_{2\pi})}{\varepsilon_{2\pi}}\right)^2 + \left(\dfrac{u(\varepsilon_S)}{\varepsilon_S}\right)^2}$

where S_{det} is the radiation-sensitive surface of the detector, and ε_S is the emission efficiency of the measured surface (expressed in radiation emitted by the surface towards the outside, in 2π sr, per radiation emitted by the contaminated surface in all directions, in 4π sr). ε_S may be difficult to determine, particularly if the radioactivity is deeply encrusted in the surface to be measured.

In this type of measurement, when appropriate, all of the radioactive contamination is measured directly at the surface, whether that radioactivity is embedded in the surface (i.e. it cannot be removed by friction) or whether it is "labile" (i.e. not bonded to the surface).

This type of "direct" surface measurement can be difficult to perform, because it is often necessary to control for the absence of surface contamination of irradiating containers or uneven surfaces. It is then preferable to take an "indirect" measurement called a "smear test" with a swab, i.e. a piece of paper, cloth or cotton that is wiped on an area S_w of the surface to be measured. This swab allows the recovery of a fraction F_{wt} of the labile radioactive contamination. To increase this fraction F_{wt}, the swab is sometimes soaked with a volatile solvent (alcohol, ...) which must be left to completely evaporate before proceeding with the measurement of radiation, particularly in case of low energy alpha and beta. The advantage of this indirect technique is that the radioactivity measurement can be performed in a location away from the sampling location, with low background radiation and flat measurement surface, which allows for efficient $\varepsilon_{2\pi}$ use.

In this case, the **labile surface activity** is calculated as:

$$A_{S|nf}[Bq/m^2] = \frac{n_{inet}[c/s]}{\varepsilon_{2\pi}\varepsilon_S I_{ray} F_{wt} S_w}$$

with $\dfrac{u_c(A_{S|nf})}{A_{S|nf}} = \sqrt{\left(\dfrac{u(n_{inet})}{n_{inet}}\right)^2 + \left(\dfrac{u(\varepsilon_{2\pi})}{\varepsilon_{2\pi}}\right)^2 + \left(\dfrac{u(\varepsilon_S)}{\varepsilon_S}\right)^2 + \left(\dfrac{u(S_w)}{S_w}\right)^2}$

where:

F_{wt} is the collection factor of the swab, i.e. the fraction of "labile" radioactivity removed by friction. This factor is often difficult to determine after only one smear test: its value is often estimated as $F_{wt} = 0.1$ without knowing its uncertainty (ISO 7503 standard).

S_w is the area wiped: 100 or 300 cm^2 is often advised, depending on the recommendations; it is difficult to estimate the uncertainty on this surface S_w, and ε_S is the emission efficiency of the radioactivity-bearing swab (expressed as outwardly emitted radiation in 2π sr per radiation emitted in all directions within the swab in 4π sr). ε_S may be difficult to determine, particularly if the radioactivity has been highly adsorbed into the swab, or if the swab is laden with thick labile material, radioactive or not.

Measuring airborne radioactivity often relies on a pump which draws a flow of air and a flat filter located between the sampling point and the pump. This device allows the collection, on the filter, of a portion of the aerosol whose size is greater than the filter's pores, or even part of the gases if the filter is loaded with an appropriate chemical reactant (this filter is often a cartridge, similar to that on a gas mask).

However, this kind of system is unable to measure noble gases such as krypton, xenon and radon. Similarly, water vapor, which can be loaded with tritium, is very difficult to measure.

To be measured, noble gases must be contained in a bottle, pressurised or not, providing that they emit X or γ radiation capable of passing throught the bottle wall.

Water vapor, and other gases can be collected by sparging through a solution.

In order to continuously measure gases that cannot be collected in a filter, ionisation chambers need to be used, in which the detection air carries the radioactive gas to be measured (see below for the use of differential chambers).

In the simple case of physical or chemical filter trapping, the volume activity of the air is given by:

$$A_{Vatm}[Bq/m^3] = \frac{n_{inet}[c/s]}{\varepsilon_{2\pi} \varepsilon_S I_{ray} F_{atm} V_p}$$

$$\text{with } \frac{u_c(A_{Vatm})}{A_{Vatm}} = \sqrt{\left(\frac{u(n_{inet})}{n_{inet}}\right)^2 + \left(\frac{u(\varepsilon_{2\pi})}{\varepsilon_{2\pi}}\right)^2 + \left(\frac{u(\varepsilon_S)}{\varepsilon_S}\right)^2 + \left(\frac{u(F_{atm})}{F_{atm}}\right)^2}$$

where:

F_{atm} is the collection factor of the filter, i.e. the fraction of atmospheric radioactivity gathered in the filter. This factor is often difficult to assess because it is highly dependent on the aerosol size and the filter's microporosity, or on the efficiency of the chemical reaction with the collected gas, the length of time required for the passage of the gas through the filter and its concentration: most often, it is evaluated for only specific scenarios, and its uncertainty is often difficult to determine.

V_p is the volume of pumped air, which can be accurately measured and whose uncertainty can be neglected.

And ε_S is the emission efficiency of the radioactivity-bearing filter (as radiation emitted in 2π sr, per radiation emitted in all directions within the filter in 4π sr). ε_S is difficult to assess depending on the rate of radioactivity adsorption in the filter and the total amount of material trapped in the filter.

The measurement itself can be deferred: the sampling filter is measured after having been exposed to the flow of contaminated air.

It can also be carried out continuously: the filter is continuously measured by a dedicated detector positioned in front of it. This filter can be either fixed or mobile. The type of device used for such continuous measurements are called "atmospheric contamination monitors".

Deducing the volume activity for continuous measurements is sometimes more difficult to exploit compared to deferred measurements, which integrate all the radioactivity over the pumping period, as does a "passive" detector.

When measuring various solid or liquid volume sources, or even filtering cartridges, which are more common than measurements performed on flat surfaces (worktop, smear or filter) in contact with the radiation detector probes, the product $\varepsilon_{2\pi} \times \varepsilon_S$ is replaced with ε, an efficiency that varies with the geometry of the source, the source material, and the position of the source relative to the detector.

ε refers either to a global counting with a detection system that does not measure the radiation's energies, or to a part of the spectrum generated with an energy spectrometry system (in this case, the selected regions of interest are the peaks that correspond to a total absorption of the energy for α and γ radiation).

Like ε also varies as $\varepsilon_{2\pi}$ and ε_S, as a function of the radiation's type and energy.

Relevant measurements are:

- laboratory measurements of various samples: earth, water, various effluents, plants, human tissue samples, etc. To facilitate measurement, these samples often require pre-treatments: calcination, dissolution, concentration, grinding, chemical separation, production of a surface deposit…;

- measurements of massive filtering cartridges or radioactive gas trapping bottles;
- measurements of humans (whole body counting, site exit portal monitors for the workers);
- measurements of vehicles (site exit portal monitors);
- in situ measurements on complex and massive matrices: drums of waste, materials from the dismantling or decommissioning of nuclear facilities, direct measurements in the environment, …

In many cases, the efficiency ε is difficult to determine because there are not always valid calibration standards which are suited to the samples to be measured, even after pretreatment. Calculations often have to be made in order to adapt the ε value to the measured sample.

Depending on whether the activity, volume activity or specific activity of the measured sample is sought, one of the following formulas is used:

$$A[\text{Bq}] = \frac{n_{\text{inet}}[\text{c/s}]}{\varepsilon \, I_{\text{ray}} \, R}$$

$$A_V[\text{Bq/m}^3] = \frac{n_{\text{inet}}[\text{c/s}]}{\varepsilon \, I_{\text{ray}} \, R \, V_S[\text{m}^3]}$$

$$A_M[\text{Bq/kg}] = \frac{n_{\text{inet}}[\text{c/s}]}{\varepsilon \, I_{\text{ray}} \, R \, m_S[\text{kg}]}$$

with:

$$\frac{u_C(A)}{A} \frac{u_C(A_V)}{A_V} = \frac{u_C(A_M)}{A_M} = \sqrt{\left(\frac{u(n_{\text{inet}})}{n_{\text{inet}}}\right)^2 + \left(\frac{u(\varepsilon)}{\varepsilon}\right)^2 + \left(\frac{u(R)}{R}\right)^2}$$

(the other terms being known quite accurately).

In these formulas:

ε is the measurement efficiency of the sample for the radiation and energy to be considered, adjusted if necessary by a correction coefficient for self attenuation, geometry, etc. ε is expressed in "counts per radiation emitted within the sample."

R is the efficiency of the physical (concentration, dilution, …) and/or chemical transformation of the initial sample.

V_S and m_S are the volume and mass of the sample before transformation.

5.3.1.7. Measurement of external exposure from the counting rate

It has been described above that the counting rate n_{inet}, adjusted for dead time and background and measured with a given detector, is proportional to the activity of a given radionuclide.

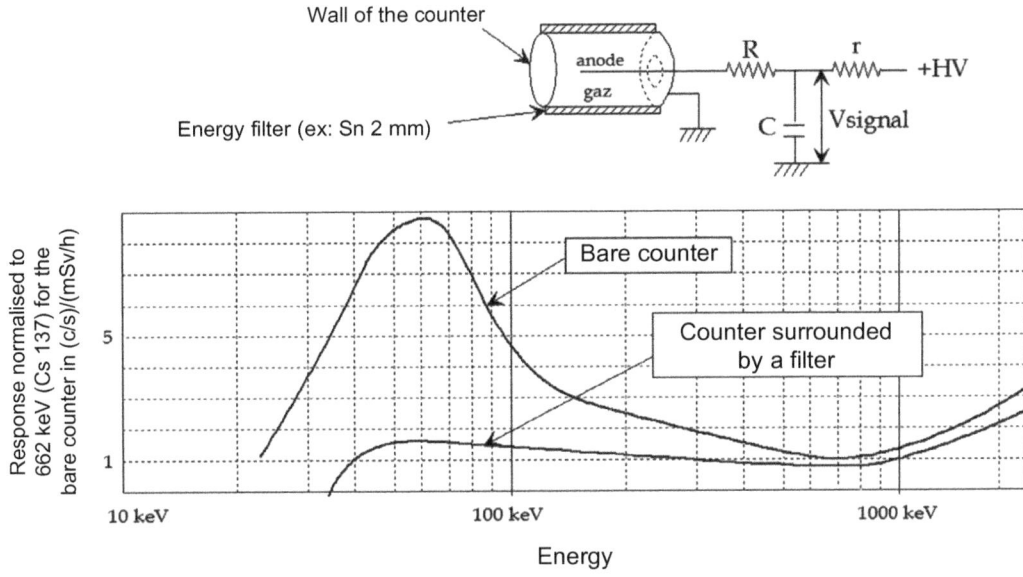

Figure 5.13. Typical response $\varepsilon_{\dot{H}}$ in (c/s)/(μSv/h) of a Geiger-Müller counter as a function of the radiation energy, with and without energy compensation filters.

Similarly, n_{inet} [c/s] is proportional to the flux of radiation of a given type and energy spectrum, and therefore to the dose-equivalent rate $\dot{H}$[μSv/h] generated by this radiation flux.

A detector such as a Geiger-Müller counter presents a response, or sensitivity, $\varepsilon_{\dot{H}}$[(c/s)/(μSv/h)] which varies greatly with the radiation energy of the photon irradiating the detector, because of the interactions in the cathode and in the detector walls.

The dose-equivalent rate $\dot{H}$[μSv/h] is calculated as:

$$\dot{H}[\mu Sv/h] = \frac{n_{inet}[c/s]}{\varepsilon_{\dot{H}}[(c/s)/(\mu Sv/h)]} \quad \text{with} \quad \frac{u_C(\dot{H})}{\dot{H}} = \sqrt{\left(\frac{u(n_{inet})}{n_{inet}}\right)^2 + \left(\frac{u(\varepsilon_{\dot{H}})}{\varepsilon_{\dot{H}}}\right)^2}$$

Detectors used to measure $\dot{H}$[μSv/h] based on n_{inet} [c/s] are usually equipped with filters that allow the reduction of variations in its response as a function of energy: such detectors are described as "energy-compensated".

The dose-equivalent rate $\dot{H}$ [μSv/h] is often relative to an operational quantity such as $H^*(10)$ (see Chapter 3, "Dosimetry").

A sample response with and without energy compensation filter is given in Figure 5.13.

The factor $\varepsilon_{\dot{H}}$ [(c/s)/(μSv/h)] is determined for a given detector in a given energy range using standard radiation beams obtained with radioactive sources or X rays generators.

The response usually varies by ± 25% in an energy range which depends on the detector, its potential energy compensation filters and its conditioning in the device in which it operates. It is increasingly common for the energy range to be shown on the device itself.

> **Further information**
>
> **Metrology, standards, calibration and metrological chain**
>
> Metrology is the science of measurement. It is based on standards and comparisons between standards of laboratories for a given quantity and radiation quality. Standards and comparisons are organised to a worldwide scale in what is called the "metrological chain". This metrological chain provides the traceability of the validity of all calibrations performed, including by the end-user of the device, by the successive calibrations.
>
> The highest metrological authority in a country is the national metrology laboratory, which is responsible for maintaining the national standards.
>
> These primary standards are sometimes derived from international standards. These international standards, as well as intercomparisons between national laboratories, are managed by the International Bureau of Weights and Measures (IBWM), which is dedicated to preserving the units of the international system.
>
> National standards give rise to secondary standards kept and managed by laboratories or metrology services that are often accredited and authorised to perform device calibration. These laboratories may offer commercial standards for device calibration by the users themselves. These commercial standards can be converted (split, diluted, packaged, ...), especially if they are in liquid form, by the users in order to create "working standards" better suited to their measurement needs.
>
> The uncertainty on the value of a quantity expressed by the standard increases with the distance from the primary standard.
>
> National standards can be dosimeters and/or reference measurement methods.
>
> When it comes to measurement methods, they are called "absolute", i.e. they intrinsically provide information about the quantity measured. These methods can also be implemented further down the metrological chain, but without any "primary" metrological value. Examples of absolute methods: measurements with a free air ionisation chamber for kerma measurement; coincidence measurements ($\gamma-\gamma$, 4π $\beta-\gamma$, etc.) for activity measurements.
>
> When a measurement method cannot provide information about the quantity to be measured until after calibration, the measurement is said to be "relative".
>
> Users can validate their calibration and, therefore, their measurement methods by participating in "inter-laboratory test programs", organised by laboratories or national technical experts. In the process of such tests, they receive samples of unknown nature and activity (for spectrometry techniques).

5.3.1.8. Measurement of internal radioactivity in a human

The measurement of internal radioactivity in a human with suspected contamination can be performed in two ways:

- Measurement of the radioactivity released by the body in the feces, urine, nose mucus, etc. Here, the measurements are carried out on the samples in a laboratory, according to the methods and precautions explained above. From these bioassay samples, the activities present in the body can be inferred thanks to a mathematical model (originating from the ICRP) of the biokinetics of the radionuclide(s) in question. In this case, assuming the samples are properly processed, nearly all of the radionuclides can be measured.

- Direct measurement of the activity present in the body, by placing the person in front of detectors that can measure the radiation emitted through the body; in practice, this means X and mostly γ radiation. This method is known as "whole body counting" or "anthropogammametry". In this case, only X- and gamma-emitting radionuclides are measurable.

The measurement efficiency ε varies from one human being to another because of the diversity in morphologies and radionuclide distribution depending on the contamination scenario.

Anthropogammametry therefore requires a specific efficiency calibration, in which the human body is simulated by an "anthropomorphic" phantom and the contamination is simulated by one or more localised or diffuse radioactive sources in the phantom. There are currently several adopted phantoms which correspond to different models and contamination scenarios.

In the vast majority of cases, the resulting activity corresponds to the calibration context and is thus expressed in "Phantom-Bq".

However, increasingly sophisticated digital corrections allow for the adjustment of the measurement result by taking into account the morphological differences between the person to be measured and the phantom used in the efficiency calibration; however, the distribution of the radionuclide in the body still remains difficult to determine with precision.

Identifying and quantifying radionuclides in anthropogammametry is based on analysing a spectrum of gamma energies which are measured in one or more large NaI(Tl)-type scintillators, which allows very brief examinations (a few minutes), or in one or more semiconductor germanium crystals, which enable spectra with narrower peaks (and thus with better energy resolution), but require longer acquisitions than with scintillators.

5.3.1.9. Detection limits in measurements

It was described in section 5.3.1.4 that, with a background leel n_b [c/s], one can define a "decision threshold" DT_α associated with the risk of "false alarm" or "false positive":

$$DT_{\alpha=2.5\%} = 2\sqrt{2n_b/t} \approx 2.83\sqrt{n_b/t} \approx 2\sqrt{n_b/\tau_{icto}}$$

$$DT_{\alpha=5\%} = 1.645\sqrt{2n_b/t} \approx 2.33\sqrt{n_b/t} \approx 1.645\sqrt{n_b/\tau_{icto}}$$

This risk α corresponds to the probability of incorrectly concluding that the adjusted counting rate n_{inet} is significant, when what is observed are unlikely fluctuations of the background.

If $n_{inet} > DT_\alpha$, it can be concluded that n_{inet} is significant to within the risk α. Its value can be retained, as well as its standard uncertainty $u(n_{inet})$, which increases as n_{inet} approaches DT_α.

$n_{inet} \pm u(n_{inet})$ is converted into activity in $A \pm u_c(A)$, which is a simple surface or volume activity, or into dose-equivalent rate $\dot{H} \pm u_c\dot{H}$ by using the factor $K \pm u_c(K)$ which corresponds to the more or less complex denominator of the formulas introduced in sections 5.3.1.5 to 5.3.1.8.

For example $K = \varepsilon_{2\pi} \varepsilon_S I_{ray} F_{atm} V_p$ for the calculation of A_{Vatm} [Bq/m³] in section 5.3.1.6.

$$A \text{ or } \dot{H} = \frac{n_{inet}}{K} \text{ and } \frac{u_c(A)}{A} \text{ or } \frac{u_c(\dot{H})}{\dot{H}} \approx \sqrt{\left(\frac{u(n_{inet})}{n_{inet}}\right)^2 + \left(\frac{u_c(K)}{K}\right)^2}$$

Since $u(n_{inet})$, $u_c(A)$ or $u_c(\dot{H})$ increases as n_{inet} approaches DT_α.

When $n_{inet} < DT_\alpha$, i.e. A or $\dot{H} < y^* = DT_\alpha/K$, it can be concluded that the measurement is not significant, i.e. it does not measure any radiation apart from the background.

However, there remains some doubt (evaluated by a risk β called risk of "non-detection" or "false negative") that the observed counting rate n_{inet} is due to a source of radiation causing an average counting rate $\bar{n}_{DL}$.

The average counting rate $\bar{n}_{DL}$, for which fluctuations around its value are observed with the low probability β below DT_α, is potentially caused by an activity or dose-equivalent rate $y^\# DL_{\beta=\alpha} = \bar{n}_{DL}/K$ called "detection limit".

The notation $DL_{\beta=\alpha}$ derives from the risk that α and β are defined as equal, which is almost always the case.

The detection limit is not easy to calculate rigourously. Its calculation is covered by the international standard ISO 11929, and is quite complex.

According to this standard, the detection limit $y^\#$ can be estimated by:

$$y^\# \approx \frac{4 + 2y^*}{1 - 4\left(\frac{u_c(K)}{K}\right)^2} \text{ for } \beta = \alpha = 2.5\% \text{ and : } y^\# \approx \frac{2.7 + 2y^*}{1 - 2.7\left(\frac{u_c(K)}{K}\right)^2} \text{ for } \beta = \alpha = 5\%$$

To better our understanding of this detection limit, we can also say that:

- If the values generated by $y^\#$ are repeated, there are β chances in 100 to observe results below $y^* = DT_\alpha/K$.

- If a measurement of $y^\#$ is attempted, there is a probability β of observing a result below $y^* = DT_\alpha/K$ and a probability $1 - \beta$ of observing a result above y^*.

The concepts of decision threshold and detection limit are illustrated in Figure 5.14.

5.3.1.10. Examples of detection systems used for activity measurements

Systems used for the measurement of surface activity

Testing for radioactive contamination of surfaces can be carried out directly, as discussed previously. To perform such measurements, a portable ratemeter (such as the one shown in Figure 5.12) is often used.

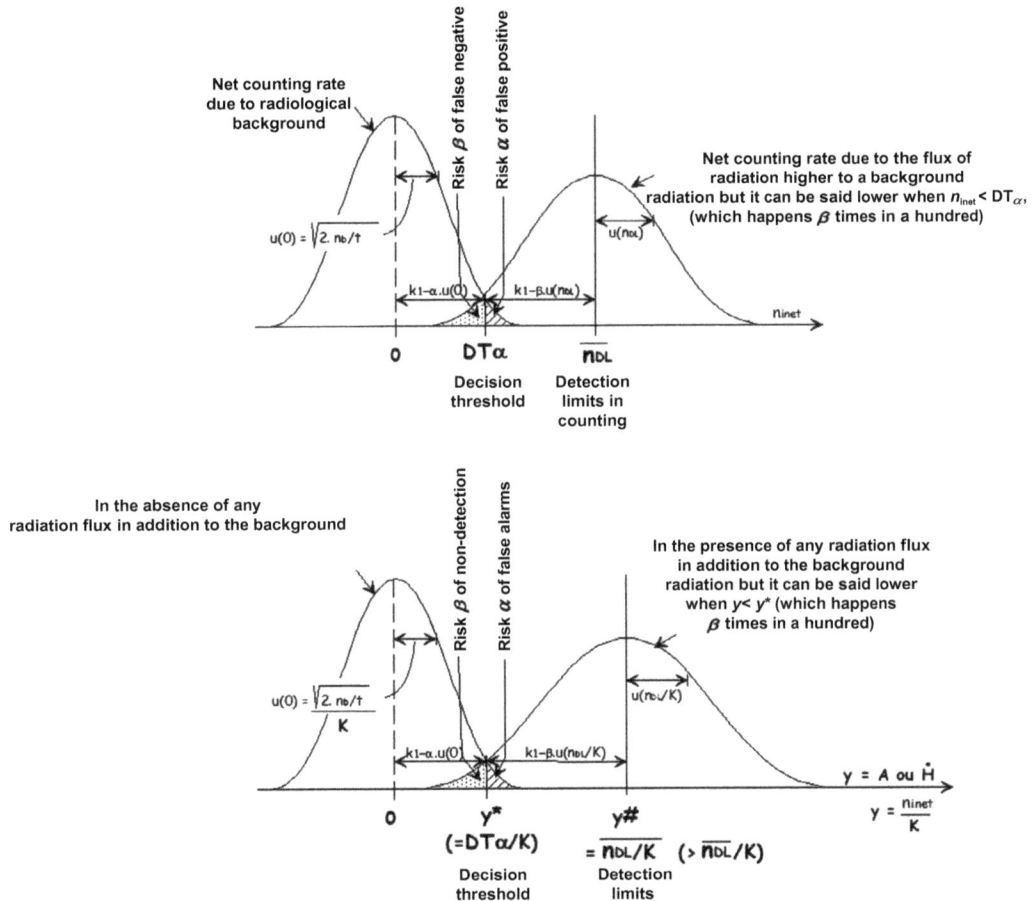

Figure 5.14. Illustration of the concepts of decision threshold and detection limit.

Probes with preamplifiers (which can incorporate the functions of High Voltage power supply), amplifier and amplitude discriminator are connected to the ratemeter unit.

The main types of detectors that can be connected to the ratemeter unit are summarised in Table 5.1.

X and gamma probes are dedicated to source search, and not to surface contamination measurements.

In addition, there are detectors known as "alpha-beta", which combine two scintillators to create a "phoswich" (contraction of "phosphor sandwich"): a deposit of zinc sulfide (ZnS) on a plastic scintillator. In this case, the alpha-beta differentiation is performed on the rise time of the detection pulses: pulses resulting from ZnS scintillations appear slower than those caused by the plastic scintillator, which appear very quickly (fluorescence time constant of about 1 ns).

In most cases of surface measurements, these detectors are fairly suitable for ground, clothing, smears, worktops, skin, etc.

Table 5.1. Probes used for contamination measurement.

Probe	Type of detector	Nature of detector	Radiation detected	Sensitivity to other radiations	Background effect (ambience ~100 nGyh)
alpha	Scintillator	Very thin ZnS on plastic substrate protected by 1 mg/cm^2 metallisation	α $E > 1$ MeV	-	~ 0.01 c/s
"soft" beta	Gas ionisation	Bell-shaped Geiger Müller counter with think mica window (1.5 to 2 mg/cm^2)	β $E_{\beta max} > 50$ keV	All others ($E_\alpha > 2.5$ MeV $E_\gamma > 10$ keV)	~ 0.5 c/s
beta	Scintillator	3 mm thick plastic scintillator protected by 25 μm metallisation	β $E_{\beta max} > 150$ keV	X et γ	~ 3 c/s
X	Scintillator	Thin NaI(Tl) (3 mm) protected by 0.2 mm of Be (40 mg/cm^2)	X and γ $E > 5$ keV	Mid-energy range β and electrons	~ 15 c/s
gamma	Scintillator	Thick NaI(Tl) (2.5 cm) protected by 0.5 mm of Al	$E_\gamma > 30$ keV	Very high energy β	~ 30 c/s

They have the drawback, however, of not providing large detection areas, which is necessary in particular for quick, comprehensive checks of hand and feet at work zone exits: because the detector must be moved on the surface to measure, there is a risk of missing small spots of radioactivity.

To carry out checks on larger areas, we must rely on proportional counters with extended surface (~ 500 cm^2) and very thin metallised windows, in which a gas, often a mixture of 90% argon (Ar) and 10% methane (CH$_4$) or CO$_2$, circulates. This economical mixture avoids attachment of the electrons generated by ionisation to the gas, and mitigates the UV radiation due to rearrangement, which triggers avalanches further in the gas (as in the Geiger-Müller counter) and disrupts the proportionality between the amplitude of the electrical detection pulse and the energy released by the ionising radiation in the gas.

The gas flow helps to prevent micro gas leaks, specifically through the thin walls and the mechanical connection to the counter's metal cathode.

This technique is utilised in hand and foot contamination monitors, head and full body skin contamination monitoring portals and mobile monitoring devices for clothing and worktops. In such devices one or more proportional counters are used, depending on the number and nature of the surfaces to be checked (one counter in a clothing monitor, 6 counters in a hand and foot monitor, and 20 counters in a monitoring portal checking for full body skin contamination).

Given their low sensitivity to X and γ radiation, and given the background (interactions in the gas only), these devices offer little to no shielding against these types of radiation.

Examples of commercial implementations of such monitoring systems are given in Figure 5.15.

Monitoring of surface contamination by indirect measurement (smear tests) is also performed with proportional counters. These can be gas flow proportional counters, such as those described in section 5.3.1.10.4. Because such devices cannot be transported to the monitoring location, we must rely on systems based on sealed counters (without gas flow) set in small, though thick, lead casing in order to exclude the ambient gamma irradiation.

Figure 5.15. Examples of large surface, gas flow proportional counters used in radiation protection for personnel, clothing and work surfaces monitoring (Source INSTN).

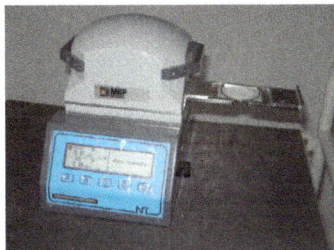

Figure 5.16. Examples of proportional counters used to control smears in indirect checks of surface radioactive contamination (Sources INSTN and Saphymo).

Twin counters are sometimes used: one counter is dedicated to the sample measurement, the other to the measurement of the background, the net measurement being the difference between the two counting rates. This twin detector technology is also implemented in differential ionisation chambers, which is discussed later in this chapter, in the context of detectors that use current measurements.

Examples of commercial implementations of such systems are given in Figure 5.16.

In wide area monitors, proportional counters are sometimes replaced by scintillators with very thin protection windows; they have the advantage of not using gas, but they have a lower detection efficiency due to the window thickness. In addition, their sensitivity to gamma radiation requires them to be shielded in order to mitigate the background radiation.

Semiconductors sometimes replace small area proportional counters in filter or smear counting applications. They offer the advantage of a good energy resolution that allows the separation not only of β detections from α detections, but also to determine the contributions from the solid decay products of the ^{222}Rn and ^{220}Rn that are naturally present in the filters (these descendants emit α radiation of energy above 6 MeV, while artificial radionuclides emit α radiation of energy below 6 MeV).

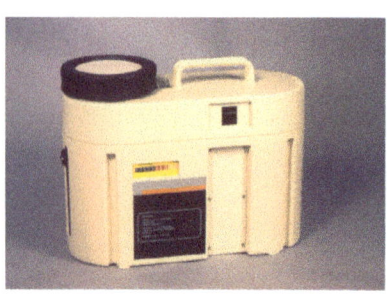

Figure 5.17. Aerosol sampling systems (Sources MGPI and INSTN).

Systems used for atmospheric activity measurements

For the measurement of aerosols and certain other gases, there are sampling devices which disclose the result after analysis of the filtering medium. Other more complex and expensive devices provide a real-time value of the volume activity. These are called atmospheric contamination monitors or beacons. These devices are available for both facility and environmental measurements.

In most cases, a distinction needs to be made between the devices performing measurements on alpha radiation (called alpha monitors) and those performing measurements on beta radiation (called beta monitors). Their sensitivities are very different, both types of monitors are often used depending on the types of radionuclides to be measured in the working atmosphere (especially in the case of alpha detection, which is much less sensitive).

Because these devices are generally very expensive, they are not very well distributed amongst the users of unsealed sources. Because regulations require the use of devices as close as possible to workstations, the most commonly used devices are fixed filter sampling monitors. Their flow rate must be representative of the workers breathing rate. They usually operate with a flow rate of 1.2 $m^3.h^{-1}$. It is possible to use devices with a higher flowrate to quickly determine if there is air contamination.

The filtering medium must be adapted to the radioactive material to be handled; for example, "coal" filters are used to measure iodines. Equipment for measuring aerosol samples is shown in Figure 5.17.

Atmospheric contamination monitors or beacons are equipped with a detector that analyses the aerosol deposit in real-time. The monitor can be connected to a radiation control panel. In general, these devices can compensate for natural radioactivity (radon), and have a higher sampling rate than the ones introduced above.

Detectors used in the monitors are either flat-panel scintillators (plastic or ZnS), proportional counters, or silicon diodes. An atmospheric contamination monitor is shown in Figure 5.18.

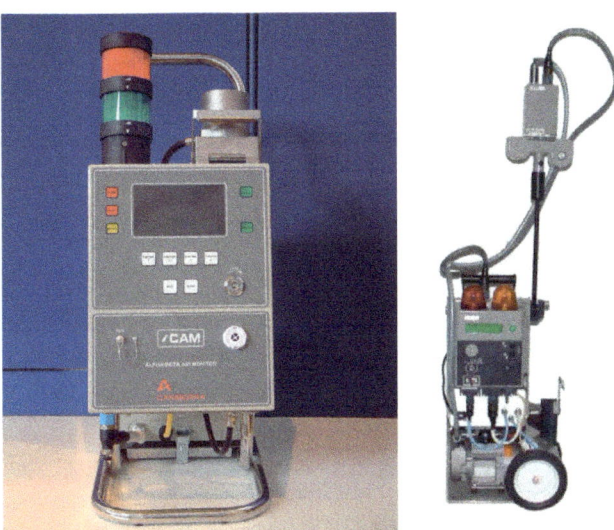

Figure 5.18. Examples of atmospheric contamination monitors or beacons (Source Canberra).

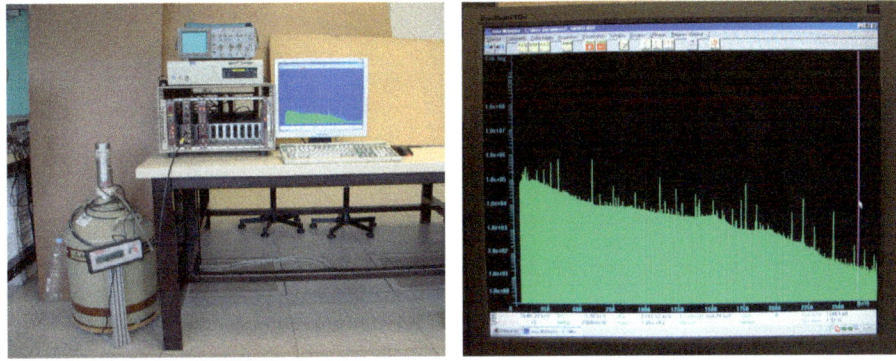

Figure 5.19. Example of gamma spectrometry chain used in laboratory measurements and typical spectrum obtained (Source INSTN).

Laboratory radioactivity measurements on various samples

Laboratory measurements mainly concern radiation spectrometry, which allows the measurement of several radionuclides at the same time, and the identification of each radionuclide and its quantitative contribution in the sample.

Laboratory measurements concern X and gamma radiation spectrometry, which is primarily performed with detection chains equipped with high purity germanium semiconductors (HPGe). An example of such a chain and the spectrum that can be obtained is given in Figure 5.19.

Laboratory measurements also concern alpha radiation spectrometry, which is performed with detection chains equipped with thin (about 100 µm) silicon semiconductor

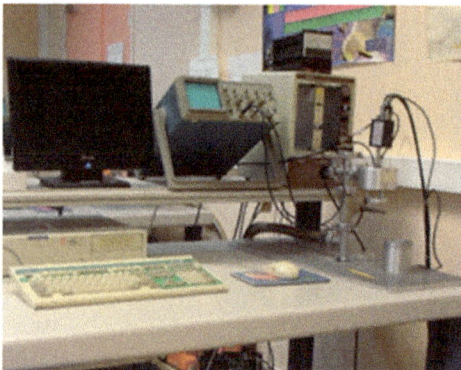 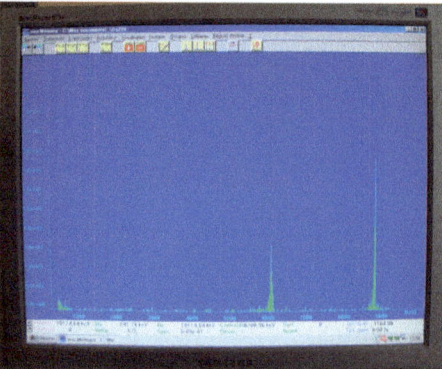

Figure 5.20. Example of alpha spectrometry chain used in laboratory measurements and typical spectrum obtained (Source INSTN).

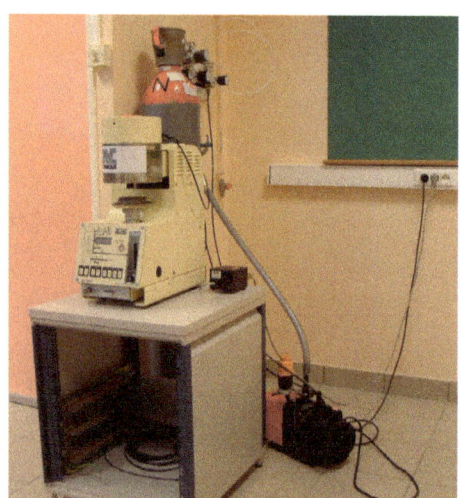 

Figure 5.21. Example of a grid ionisation chamber (Source INSTN).

placed in a rough vacuum chamber. An example of such chain and the spectrum that can be obtained is given in Figure 5.20.

Alpha spectrometry measurements can also be obtained with a particular ionisation chamber working in pulsed mode, called a "grid chamber". An example of industrial application is given in Figure 5.21.

Beta radiation spectrometry and internal conversion electron spectrometry can be performed with silicon semiconductor detectors of greater thickness than for alpha radiation although this technique is used infrequently.

When electrons' energy is low (very soft β such as those of tritium ^{3}H), measurements with liquid scintillation (explained earlier in this chapter) are performed using equipments involving automatic sample changers, as illustrated in Figure 5.22. This technique also allows the measurement of low level alpha radioactivity (PERALS method) (Source INSTN).

 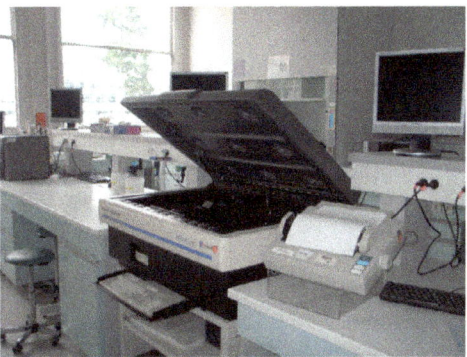

Figure 5.22. Examples of commercial implementations of a liquid scintillation spectrometer (Source INSTN).

The spectrometry techniques described above are implemented with multi-channel analysers. Single-channel devices or a simple discriminator allow simple selections within a spectrum. This system is often used in measurements performed with well-shaped scintillation crystals, into which the sample is introduced. As with liquid scintillation, the well-detectors allow excellent measurement efficiency thanks to the way the detector envelops the sample, as shown in Figure 5.23.

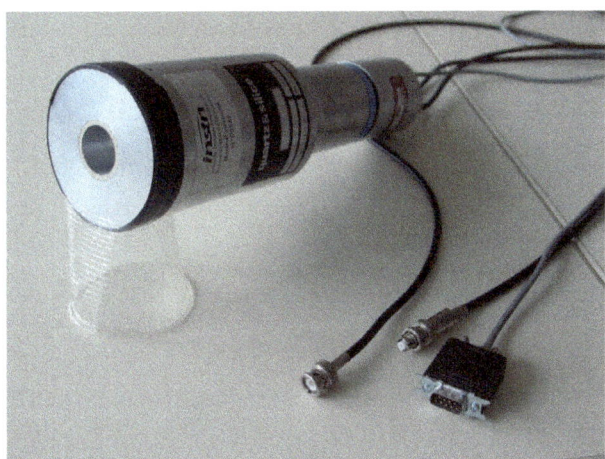

Figure 5.23. Example of scintillation well (Source INSTN).

The counting of low level filters and various samples (smears, aerosol samples, charred handkerchiefs on a dish, ...) can also be performed with gas flow proportional counters (or even Geiger Müller counters) with very thin windows and protected by substantial thicknesses of low natural residual radioactivity lead.

This type of detector can measure alpha and beta radiation at the same time: the counting is thus called "total $\alpha - \beta$". α and β radiation can also be measured separately by amplitude discrimination of the pulses generated: the highest amplitudes correspond to the high energy deposited by alpha radiation.

Large thicknesses of lead reduce background to very low levels, e.g. 40 counts per hour for β detection and 2 c/h for α detections.

Multiple detectors within the same device allow for parallel measurement of multiple samples.

An example of such counting system is given in Figure 5.24.

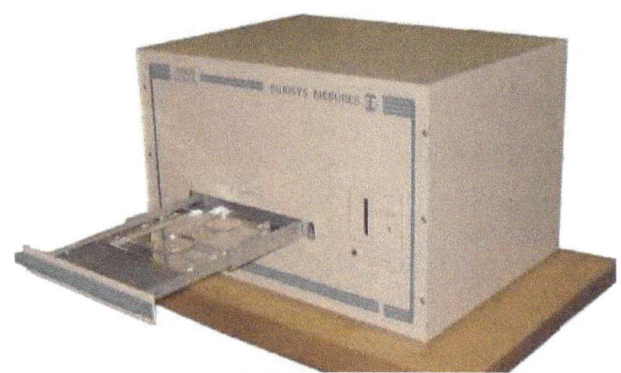

Figure 5.24. Example of a counting system with low background for the measurement of alpha and beta radiation (Source Canberra).

Systems used for on-site assessment of the global radioactivity of objects, people and vehicles

To be sure that workers are not contaminated when exiting the working area, and to prevent the theft of radioactive material, portal monitors are often placed at the exits of a nuclear facility.

These portals are usually sensitive only to gamma radiation. In order to optimise measurement efficiency, such portals use several large, thick (several centimeters) solid state detectors positioned on all sides of a compartment surrounding the person. Measurements take only a few seconds each.

Plastic scintillators are used because they are easily manufactured as panels measuring 50 cm × 50 cm or larger. They have a rapid fluorescence, which allows detection of very rapid changes in the counting rate above the background radiation.

The scintillators are often partially protected from the background by a few cm of lead on the outer face of the portal.

An example of commercial portal is shown in Figure 5.25. This type of portal allows detection of 3 kBq of ^{60}Co activity in 1 s.

This type of detection is also used to control vehicles and their cargo. An example of this is illustrated in Figure 5.26.

This type of detection with large volume plastic scintillators is used for spot checks. An example of a very simple system of this type is shown in Figure 5.27.

Inorganic scintillators such as NaI(Tl) are occasionally used in the manufacture of vehicle and cargo monitoring portals. They are often found in lead-shielded casing into which "small objects" are introduced in order to measure their emission of gamma radiation.

Similarly, at the exit of a controlled zone with a high risk of contamination by gamma emitting radionuclides (such as cobalt 60 in nuclear reactors), there are control portals

Figure 5.25. Example of personnel gamma radiation detection portal monitor at the exit of a nuclear facility (Source Canberra).

Figure 5.26. Example of vehicle radiation portal monitor at the exit of a nuclear facility (Source Saphymo).

based on NaI(Tl) scintillators for personnel wearing working clothes. Because they are highly sensitive to the background's gamma radiation despite their lead shielding, they have a high detection limit (about 7 kBq ^{60}Co).

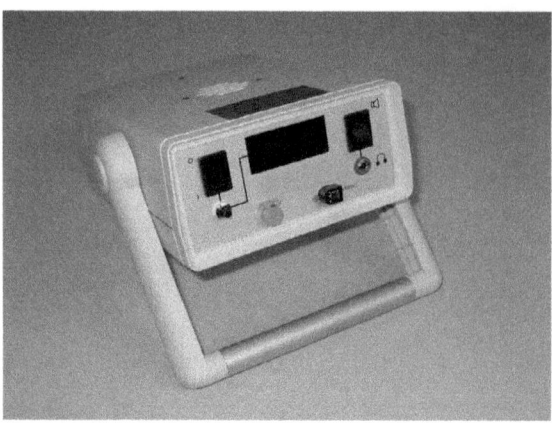

Figure 5.27. Example of a high efficiency portable gamma detection system for radiation spot checks (Source Saphymo).

Systems used in whole body counting

The systems mentioned above are not used to perform spectrometry because the low average atomic number of organic scintillators does not allow the total absorption of the gamma radiation's energy and does not generate peaks in the spectra. Additionally, their detection limit is usually high, typically at least one kBq. Indeed, the background radiation is only slightly mitigated by the lead shielding which, when present, is only of moderate thickness.

Furthermore, regarding NaI(Tl) scintillators, the volume of the detectors used does not allow adequate total energy absorption efficiency for high energy gamma rays.

We instead turn to large volume NaI(Tl) scintillators (several litres), singly or in pairs, to scan the body. The scintillators may be fixed or mobile. They, as well as the person to be measured, are shielded from the background radiation by very thick layers of lead or steel with very low residual radioactivity. Consequently, whole body counters weigh several tons and require a special infrastructure, such as compartments (with thick shielded walls) or sarcophagi. A commercial whole body counter is illustrated in Figure 5.28, along with a model of an anthropomorphic phantom used for efficiency calibration.

Measurements with this type of system take at most a few minutes, with the person being measured in standing position. Activities of 500 Bq are easily shown in one minute of acquisition for the most common products of fission and neutron activation.

Spectrometric systems with higher resolution are used under similar conditions with HPGe semiconductor detectors. As it is not possible to manufacture large volumes of these detectors, efficiencies are also lower and examinations require at least ten minutes (often more), which requires the person to be measured to sit or lie, which is also the case when small volume scintillators are used.

5.3.1.11. *Examples of detection systems used for external exposure measurements*

It was described in section 5.3.1.8 that the measurement of photon-induced external exposure could be deduced from the counting rate, after calibration.

Figure 5.28. Example of whole body counting system (weight: about 5 tons) and anthropomorphic phantom used for calibration (Sources Canberra and INSTN).

Many radiation meters, based on Geiger-Müller counters and with energy compensation, work on this principle. These counters are small enough not to saturate at high levels of irradiation. They are cylindrical instead of bell-shaped with a thin window, a configuration which is used to measure surface radioactivity.

The meters used are small so that the counting does not saturate with the level of irradiation. Their small size does not require the application of a high polarisation voltage, which is thus easy to generate for portable devices.

Some industrial examples of "pocket" radiation meters are shown in Figure 5.29.

Some radiation meters include a proportional counter, also equipped with energy-compensation filter(s), instead of a Geiger-Müller counter. This option allows the reduction of losses due to dead time. This is specifically the case with the FH 40-type radiation meter marketed by Thermo Scientific.

Radiation meters are also used as gamma monitors placed on mobile sites. Figure 5.30 shows a few examples of gamma monitors based on Geiger-Müller counters.

This principle of converting counting rates into dose-equivalent rate is also applicable to the measurement of neutron induced irradiation. The detectors used are proportional counters in which a converting material is present either as a detection gas (for ^{3}He or ^{10}BF$_3$ counters) or a thin inner wall on the counter's cathode (for counters with boron deposits highly enriched with boron-10).

Nuclear reactions converting neutrons to charged particles capable of ionising the gas are ^{3}He$(n, p)^3$H and ^{10}B$(n, \alpha)^7$Li.

These conversion reactions are very favourable only in the case of thermal and epithermal neutrons. Therefore counters are placed at the centre of substantial volumes of highly hydrogenated materials such as polyethylene, which can thermalise the flux of neutrons, thus making them detectable. The volume of peripheral polyethylene can be varied according to the energy of the neutrons to be measured. In systems with multiple detectors, cadmium is occasionally used to absorb thermal neutrons and measure only fast neutrons.

 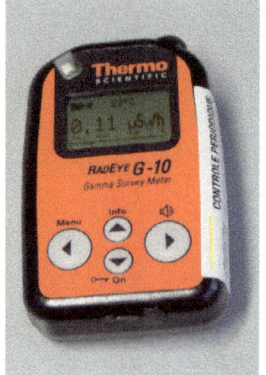 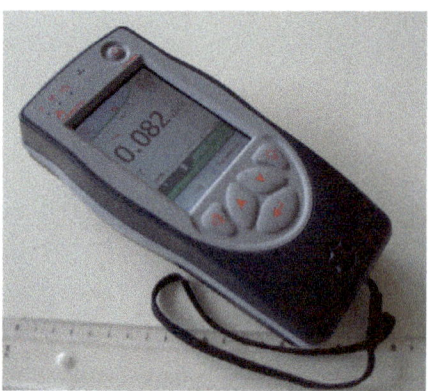

Figure 5.29. Examples of "pocket" radiation meters incorporating an energy-compensated Geiger-Müller counter and based on the conversion of a counting rate into a dose-equivalent rate (Source INSTN).

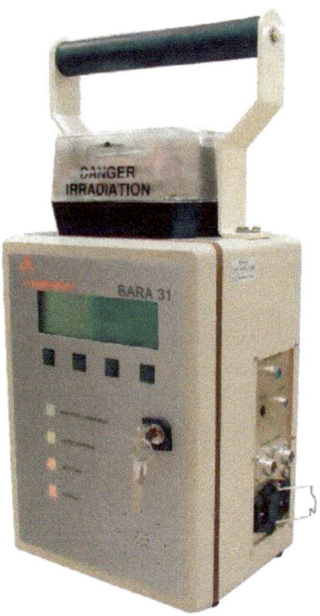 

Figure 5.30. Examples of gamma radiation monitors based on Geiger-Müller counters (Sources Canberra and Berthold).

Some examples of neutron radiation meters are given in Figure 5.31.

Some devices also include a scintillation detector such as an NaI(Tl) or even a Cd(Zn)Te semiconductor detector to simultaneously perform gamma spectrometry screening of the radionuclides responsible for irradiation.

The counting rates of scintillation or semiconductor detectors can be converted to dose equivalent rate with or without the associated Geiger-Müller counter. This is the case with

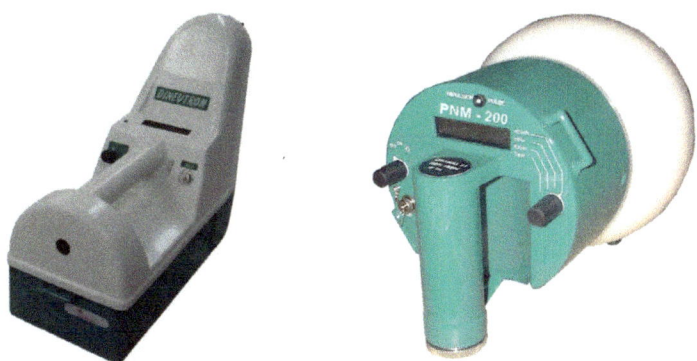

Figure 5.31. Examples of neutron radiation meters (Source Canberra).

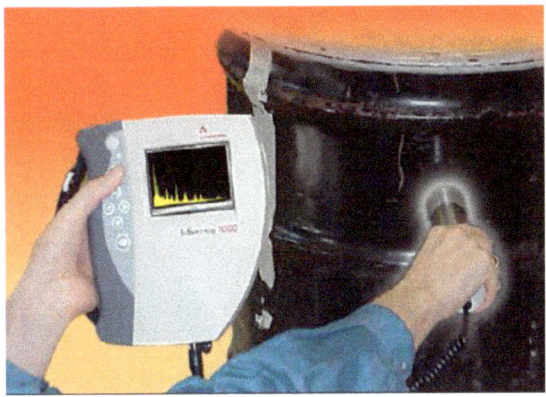

Figure 5.32. Example of radiation meter with solid state detectors, capable of gamma spectrometry screening (Source Canberra).

the detectors shown in Figure 5.32, which are also often capable of identification by gamma spectrometry.

Finally, detectors that take the form of silicon semiconductor pellets are used for the measurement of dose equivalent rates. They are found mainly in operational dosimeters capable of measuring and integrating rates of external exposure.

The detectors' small volume provides for a very small conversion efficiency ε_H: about 0.035 c/s per µSv/h, i.e. 130 c/µSv. Therefore, they allow easy measurement of doses of one tenth of µSv.

The use of these operational dosimeters is mandatory for radiation protection in certain restricted areas, and readings occur both on the device and through terminals linked to a dose management computer network. They allow optimisation of the doses received during operations in an irradiating area, and have adjustable alarm thresholds to warn, with an alarm noise, the user performing his/her work, often in a specialised protective suit.

The detectors are energy-compensated by a set of filters to obtain a nearly flat response as a function of the photons' energy.

They can also be used to monitor the dose equivalents due to neutrons using a set of converters and moderators.

An operational dosimeter thus often integrates a set of small silicon diodes and a set of compensating filter or converting material layers.

These dosimeters display a dose equivalent relative to the operational quantity $H_p(10)$ (see Chapter 3, "Dosimetry").

Thin silicon does not require application of high polarisation voltages (a few dozens of volts), which simplifies the design of these highly portable devices.

Some examples of industrial implementations are shown in Figure 5.33.

Figure 5.33. Examples of operational dosimeters (Sources INSTN and Mirion).

5.3.2. *Measurement of an ionisation current*

Measuring counting rates is not the only way to perform measurements; we can also measure average ionisation currents that are proportional to the irradiation level of the gas in the detector to be used, which in this case is an ionisation chamber.

5.3.2.1. *Measuring external exposure*

The operating voltage of an ionisation chamber ranges from 60 to 300 V, depending on the size of the chamber, whose shape can be cylindrical or planar.

The chambers may be sealed or open to the surrounding environment, and can integrate the charge or measure the average ionisation current.

The filling gas, which may be air, behaves almost identically to human tissues with regards to directly ionising charged radiation such as electrons. The dose absorbed by these tissues can be inferred from the dose absorbed into the gas by multiplying by a coefficient whose value is close to one and almost independent of the incident radiation's energy.

In the case of electromagnetic radiation, interactions in the gas are negligible compared to the interactions in the detector's wall. The electrons set in motion by the electromagnetic radiation in the wall are the ones which ionise the gas.

This fact is associated with the "Bragg-Gray" principle through which currents measured by gas ionisation chambers can be effectively used to measure the absorbed dose.

According to this principle, which applies only to X and γ photon dosimetry, we can calculate the dose D_m absorbed in a given medium m as a result of the number N_{pi} of ionisations generated in the gas contained in a small cavity present in this medium:

$$D_m = N_{pi} \varpi_{gas} \frac{S_m/\rho_m}{S_{gaz}/\rho_{gaz}}$$

where: ϖ_{gas} is the average energy required to induce an ionisation in the gas;
S is the stopping power (or linear energy transfer) by collision of the electrons set in motion by incident photons.

This principle is illustrated in Figure 5.34.

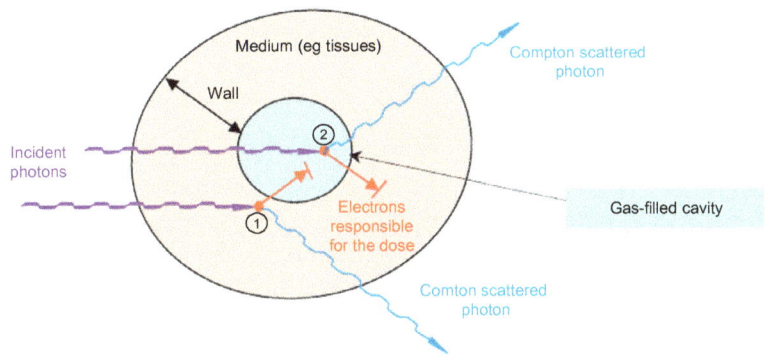

Figure 5.34. Illustration of the Bragg-Gray principle. The scopes of the electrons are represented in red scenarios such as ① tend to balance with scenarios such as ②.

As with all other detectors, precautions should be taken with ionisation chambers since in cases where the gas is not uniformly exposed to the radiation field, the measurement is going to be underestimated. This is particularly pronounced in the case of ionisation chambers, as they require significant volume in order to be sensitive to low levels of radiation, because the charges are not multiplied in the gas.

Ionisation chambers are used in fixed installations and are often connected to the radiation control panel.

They can also take the form of portable devices which include an electrometer.

A well known example of a portable ionisation chamber is the "Babyline" marketed by Canberra.

The Babyline's ionisation chamber has a volume of approximately 500 cm³, containing air at atmospheric pressure. The wall is composed of material which is equivalent to human tissue in terms of atomic composition and density. A removable cover allows the measurement of the dose absorbed in the tissues to a depth of 70 μm (thickness of 7 mg.cm⁻² which corresponds to the depth of the stratum corneum) or to a depth of 3 mm (thickness of 300 mg.cm⁻² which corresponds to the depth of tissue at which the former quantity $Dt(3)$ need to be measured but used for $H^*(10)$ measurement under well suited calibration).

The measuring range of the tissue-absorbed dose rate lies between 5 μGy.h⁻¹ and about 100 mGy.h⁻¹.

This device, along with others, is shown in Figure 5.35.

Figure 5.35. Examples of ionisation chambers used in radiation protection to measure external exposure due to electromagnetic radiation (Source INSTN).

5.3.2.2. Measurement of radioactive gases

Special ionisation chambers known as differential chambers are used for gases that cannot be trapped in filters. The gas to be measured is introduced, continuously or not, in one chamber; then the ionisation current is compared with the other chamber, which is identical to the first, except that no gas has been introduced. The second chamber, located next to the first one, measures the ambient exposure, and the difference between both ionisation currents gives the level of radioactivity of the gas in the chamber, after an often difficult calibration.

These devices are primarily used to measure noble gases and tritium in gaseous form. They can be connected to a radiation control panel.

5.3.3. Integrating ionisations over the duration of exposure: passive detectors

These detectors can only measure an absorbed dose. Their stored data can be accessed only after the end of the exposure, using appropriate reading devices.

These passive dosimeters are called "latent effect", or "delayed reading" dosimeters.

5.3.3.1. Measuring external exposure by photographic dosimetry

This dosimeter was widely use until fairly recently. It helped to record doses received by workers at skin and whole body levels. Until the end of 1998, it was the dosimeter imposed by regulations in France.

The PS1 dosimeter, distributed by the *Institut de Radioprotection et de Sûreté Nucléaire* ("Radioprotection and Nuclear Safety Institute"), is illustrated in Figure 5.36, and an exploded view is provided in Figure 5.37.

Because a given photographic film only covers a narrow range of doses and has very heterogeneous responses to energy, this dosimeter is technologically sophisticated. Several different sensitivity ranges (according to the size of the AgBr grains) allow measurements

Figure 5.36. PS1 dosimeter (Source CEA).

Areasscreens
A - Bare area
B - 300 mg.cm^{-2} plastic
C - 1.5 mm aluminium
D - 0.2 mm copper + 1.3 mm aluminium
E - 0.2 mm copper + 0.4 mm copper
F - 0.34 mm cadmium + 0.6 mm tin + 0.4 mm lead
G - 1 mm tin + 0.4 mm lead

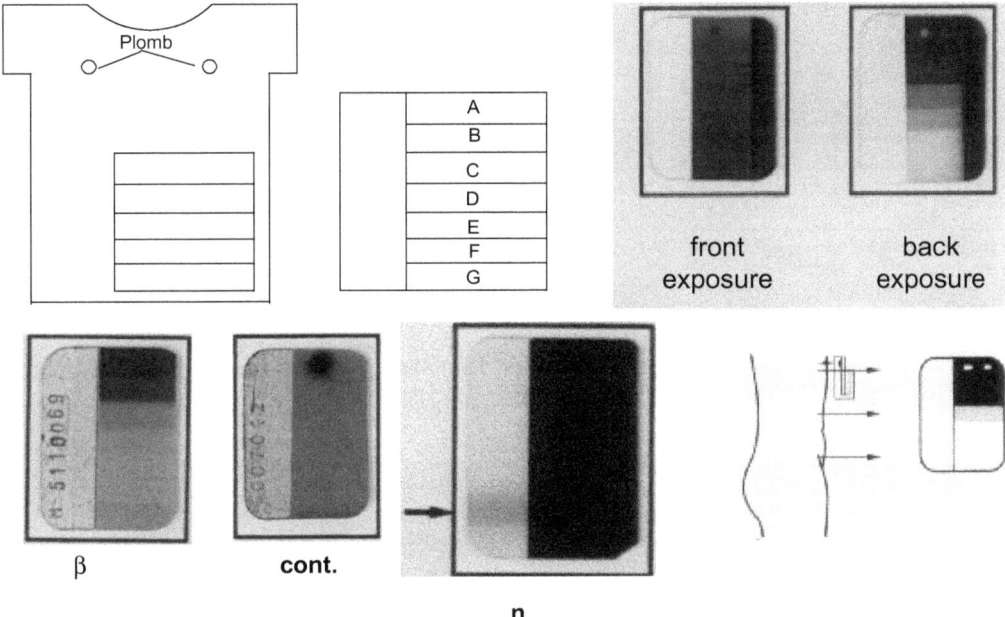

Figure 5.37. Exploded view of the PS1 dosimeter film (source CEA) which allows to clearly see the complex set of energy compensating filters.

of a wider range of doses. The casing, which approximates human tissue, incorporates an elaborate set of filters that allow to compensate the over-response at low photon energies, and to obtain a crude approximation of the energy spectrum of the radiation field which caused the exposure being recorded. The screenless part also allows the measurement of high-energy beta radiation.

This detector retains an image of the dose received, which is advantageous, especially in case of posterior inquiry.

5.3.3.2. *Measuring external exposure by TLD*

The radio-thermo-luminescent detectors (TLD) detector under incidence of X, γ, β radiation and charged particles allows the measurement of the operational quantity, $H_p(0.07)$.

Their small volume (a few cubic millimeters) enables them to perform measurements such as the dose absorbed "at contact", specifically monitoring the exposure of the operators' fingers.

The dosimeter is reusable after reading, but the information contained is destroyed, as it is partially converted into light energy. However, a second reading remains possible by acting on deeper traps, solicited at higher temperatures. Figure 5.38 shows some examples of dosimeter pellets and badges.

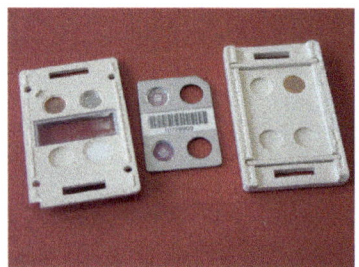  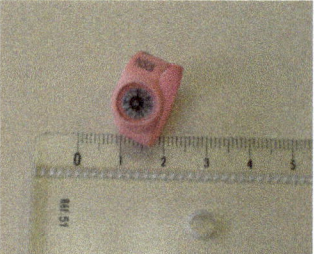

Figure 5.38. Examples of radio-thermo-luminescent dosimeter pellets, rings and badges (Source INSTN).

5.3.3.3. *Measuring external exposure by OSL*

Optically stimulated luminescence (OSL) dosimeters also appear as more or less complex badges.

The OSL detector, under the incidence of X, γ, β radiation and charged particles, allows the measurement of the operational quantities $H_p(10)$, $H_p(0,07)$ and $H^*(10)$ (see Chapter 3, "Dosimetry").

A commercial example is given in Figure 5.39.

5.3.3.4. *Mesuring external exposure by RPL*

Radio-photo-luminescence (RPL) dosimeters also take the form of more or less complex badges. A commercial example is given in Figure 5.40 below.

As the OSL, the RPL allows the measurement of the operational quantities. $H_p(10)$, $H_p(0.07)$ and $H^*(10)$ (see Chapter 3, "Dosimetry").

Figure 5.39. Example of OSL dosimeter badge (Source LCIE LANDAUER).

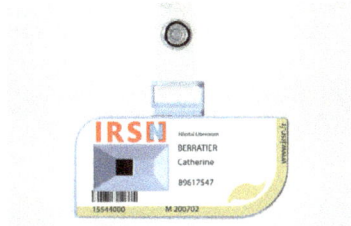

Figure 5.40. Radio-photo-luminescent dosimeter badge (Source IRSN).

5.3.3.5. *Measuring external exposure to neutrons with solid state track detectors*

In addition to measuring the dose equivalents produced by RPL and OSL badges for photons, measuring neutrons requires a solid state track detector to be incorporated into these badges, usually behind one or more converters, to enable exploration of a wider range of neutron energies.

Figure 5.41 illustrates some explanations and commercial implementations of this process.

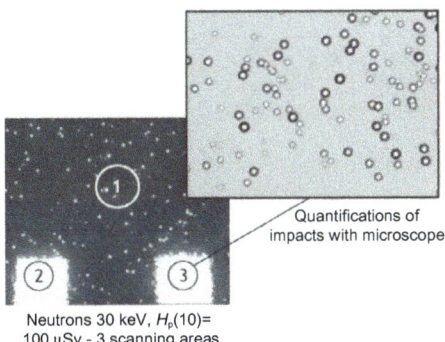

Figure 5.41. Solid state track detectors implemented for neutron detection.

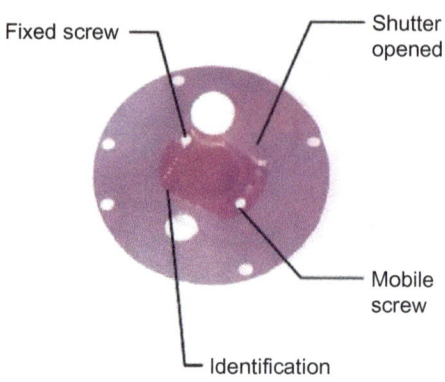

Figure 5.42. Solid state track detectors used for the measurement of atmospheric alpha radioactivity (Source Algade).

5.3.3.6. Measure atmospheric alpha radioactivity with solid state track detectors

Measuring the atmospheric alpha radioactivity over long periods of exposure often relies on allowing potentially alpha radioactivity-laden air to dwell (statically or dynamically) in the vicinity of a solid state track detector.

The natural radioactivity caused by ^{220}Rn and ^{222}Rn fumes is particularly suited to measurement by this technique, in order to monitor the atmosphere in housing as well as the exposure of uranium and thorium miners.

To monitor the air in housing or in the environment, a vessel containing a solid state track detector film is turned over at ground level.

Monitoring miners utilises a dynamic sampler with an aspiration flow representative of the miners' breathing; the air thus drawn is blown in the vicinity of a solid state track detector film. This measurement allows the deduction of the internal exposure that the minors are subject to, external exposure being separately measured by a TLD badge, often set in the same device.

Figure 5.42 shows some examples of implementations.

5.3.3.7. Measuring intense neutron fluxes by activations

Assemblies of materials which can be activated by neutrons with variable, energy-related cross sections are made in order to produce criticality pellets or belts.

Belts are worn by personnel who may be exposed to a rapid neutron excursion during accidents, such as Tokai Mura (September 1999 - Japan).

The pellets are arranged in a fixed configuration and must be carried by personnel during evacuations when a neutron excursion is detected by dedicated monitors.

In all cases, the measurement is performed after exposure and allows an estimate of the accidental dose equivalent.

Figure 5.43 illustrates such detectors.

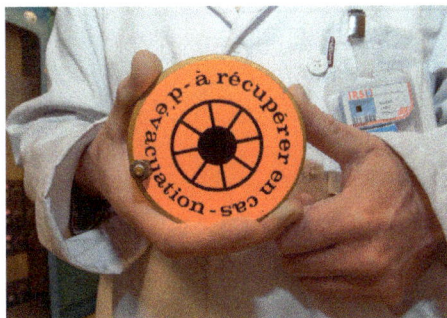

Figure 5.43. Examples of criticality dosimeters.

5.4. Check your Knowledge

1. **The scintillation light due to the interaction of one radiation: is it visible with the naked eye?**

   Answer: No. The amount of light is too weak, even in total darkness and after visual accommodation. If a high LET particle interacts into a scintillator, the resulting scintillation spark can be seen with a magnifying lens (like in Rutherford's experiment or in a spinthariscope). In the other cases a very sensible photoelectric device is needed to detect one scintillation flash.

   However, the sum of all the scintillations generated at the same time by a large amount of radiations in a fluorescent screen is easily visible with the naked eye.

2. **As its name suggests, does the photomultiplier tube, coupled to a scintillator, multiply the number of light photons emitted by the scintillator?**

   Answer: No, it first converts a part of the number of scintillation photons into electrons within the photocathode. Then this number of photo-electrons is multiplied along the dynodes by a huge factor. So a photomultiplier tube multiplies the electrons generated by the scintillation photons.

3. **When the light signal delivered by a scintillator is very weak, which kind of experimental device is necessary for separating the light scintillations from the thermal noise of the photomultiplier tube?**

   Answer: The use of 2 photomultiplier tubes is necessary. The noise of each photomultiplier tube gives weak pulses at random. Consequently the noise pulses from each tube are not coincident in time. On the contrary, a scintillation flash gives rise to 2 pulses, one on each PM tube, at the same time. A coincidence unit validates only the scintillation pulses which are then summed because of their weak height. Nevertheless, the noise of the 2 PM tubes can generate some random coincidences.

4. **How many electrons are released on average in a gas when a radiation transfers 60 keV of its kinetic energy? Will the electric pulse resulting from this interaction be easily measurable?**

    Answer: Into a gas, on average, 30 eV are needed to create an electron-ion pair. Thus 100 keV create: 60 000 eV / 30 eV / e-ion-pair = 2000 electrons released (and also 2000 ions).

    If we consider a very small capacitor of 1 pF, 2000 electrons stored into this capacitor generate a voltage of: (2000 electrons × 1.6 10^{-19} C/electron) / 10^{-12} F = 3.2 10^{-4} V = 0.3 mV.

    It is quite impossible to extract this pulse from the electronic noise. Consequently more electrons must be generated. The High Voltage applied to the gas tube must be increased to cause Townsend's avalanches and multiply the number of primary electrons by a factor of many orders of magnitude (Proportional counter or Geiger counter).

5. **What do we measure with an ion chamber like the "Babyline":**

    a) **with its 300 mg.cm^{-2} cover?**

    b) **without its 300 mg.cm^{-2} cover?**

    Answer: a) $H^*(10)$ through a suitable calibration.

    b) The absorbed dose rate at a depth of 70 μm (thickness corresponding to the horny layer of the epidermis).

6. **Give the order of magnitude of the dead time τ of a Geiger-Müller counter? If a Geiger tube generates a count rate r_c = 1000 counts per second, calculate r_i the interaction rate into the gas of the tube.**

    Answer: ~10^{-4} s or 0.1 ms

    With a dead time of 0.1 ms, the fraction of the time where the counter is available for detection (because not busy with the creation of pulses) is: $1 - r_c.\tau$ = 1 − (1000 s^{-1} × 10^{-4} s) = 0.9 = 90%

    It means that 10% of time, the counter is "dead" for detection.

    Thus, on average, the number of interactions during each second is: r_i =1000 s^{-1}/0.9 ≈ 1111 s^{-1}.

    The rate of counting losses is: 1111 − 1000 ≈ 111 s^{-1}. (111/1000 = 11% of the interactions do not give rise to a pulse)

    In this calculation we assumed that an interaction occurring during dead time does not change the duration of this dead time.

7. **Give an example of a case in which an ion chamber like the "Babyline" provides an erroneous value by underestimating the absorbed dose rate?**

    Answer: In all cases where the sensitive volume of the ionisation chamber is not uniformly bathed by the radiation field (for example, narrow radiation beams). Another case is when the energy of radiation is lower than the bottom value of the energy response.

8. **What is the threshold value above which equivalent doses received for common dosimeter films are measured?**

    Answer: $H^*(10) = 0.20$ mSv

9. **Is the response of photographic emulsion independent from the radiation's energy?**

    Answer: No (a matrix of shields is necessary to flatten its energy response).

10. **Name the major latent reading dosimeters.**

    Answer: Dosimeter film – radio-thermo-luminescent dosimeters (TLD) – radio-photo-luminescent dosimeters – optically stimulated luminescent dosimeters (OSL) – track-etch detectors – activation detectors for neutrons.

11. **What is the purpose of the filters on the casing of dosimeter films?**

    Answer: They allow analysis of the radiation field (type and energy of the incident radiation) to which the film was irradiated.

12. **What is indicated by the term "direct reading dosimeter"? Give an example.**

    Answer: A dosimeter is called "direct reading" when the detector/reader coupling is achieved. A dosimeter pen is an example of a direct reading dosimeter.

13. **What is the difference between a direct reading and a latent reading dosimeter?**

    Answer: Direct reading dosimeters immediately give the information. Latent reading dosimeters need to be evaluated before the information can be obtained.

14. **Which type of dosimeter should be used to obtain the most accurate measurement of the contact dose?**

    Answer: Radio-thermo-luminescent dosimeters.

    (Lithium fluoride pellets often called, FLi).

15. **What are radio-thermo-luminescent dosimeters used for?**

    Answer: They allow measurement of the dose absorbed through "contact" with a radioactive source. They also allow a more precise dosimetry of the equivalent dose received by the extremities, and they are increasingly used for passive dosimetry of the whole body.

16. **What type of device should be used for control in the event of surface contamination of an unknown nature?**

    Answer: A device with multiple detector probes, that can control each type of radiation.

17. **Electronic dosimeters are individually assigned. From which value is the dose equivalent generally recorded?**

    a) 100 µSv

    b) 10 µSv

    c) 1 µSv

    d) 0.1 µSv

    Answer : c) 1 µSv

18. During $\Delta t = 10$ seconds, the gross counting of a radioactive contamination spot is: $n_{gross} = 144$ counts. Estimate the standard uncertainty on this counting.

    The counting of the background radiation around this measurement is $n_{bg} = 25$ counts during $\Delta t = 10$ s.

    Calculate r_{net}, the net counting rate due to the radioactive spot and estimate its standard uncertainty.

    Answer: $u(n_{gross}) = \sqrt{n_{gross}} = \sqrt{144} = 12$ counts

    $n_{bg} = 25$ and $u(n_{bg}) = \sqrt{n_{bg}} = \sqrt{25} = 5$ counts

    $n_{net} = n_{gross} - n_{bg} = 144 - 25 = 119$ counts

    $u_c(n_{net}) = \sqrt{u^2(n_{gross}) + u^2(n_{bg})} = \sqrt{n_{gross} + n_{bg}} = \sqrt{144 + 25} = \sqrt{169} = 13$ counts

    $r_{net} = n_{net}/\Delta t = 119$ counts/10 s $= 11.9$ c/s

    $u_c(r_{net}) = u(n_{net})/\Delta t = 13$ counts / 10 s $= 1.3$ c/s

    Finally: $r_{net} = (11.9 \pm 1.3)$ c/s

    (About 2 thirds of the results of future measurements should occur between 10.3 and 13.2 c/s).

19. If the counting rates of question 18 were measured by a counting rate meter with a time constant $\tau = 2$ seconds, calculate the standard uncertainty on r_{net}.

    Answer: $r_{gross} = 14.4$ c/s and $u(r_{gross}) = \sqrt{r_{gross}/(2 \times \tau)} = \sqrt{14.4 \text{ s}^{-1}/2 \times 2\text{s}} \approx 1.9$ c/s

    $r_{bg} = 2.5$ c/s and $u(r_{bg}) = \sqrt{r_{bg}/(2 \times \tau)} = \sqrt{2.5 \text{ s}^{-1}/2 \times 2\text{s}} \approx 0.8$ c/s

    $u_c(r_{net}) = \sqrt{u^2(r_{gross}) + u^2(r_{bg})} = \sqrt{(1.9 \text{ s}^{-1})^2 + (0.8 \text{ s}^{-1})^2} \approx 2.1$ c/s

    Finally: $r_{net} = (11.9 \pm 2.1)$ c/s.

    The measurement is less accurate.

20. Is it possible to measure ^{14}C inside the body of a worker with a conventional whole body counting device?

    Answer: No. ^{14}C is a pure beta decay radionuclide. No gamma radiation is emitted. Endpoint energy of ^{14}C beta is 156 keV. The Beta ray energy is not high enough to be transmitted through the body.

6 Uses of sources of ionising radiation

Cécile Étard

Introduction

Sources of ionising radiation are used widely in medical and industrial fields as X-ray generators, accelerators, sealed or unsealed radioactive sources.

The most numerous and best known applications relate to the field of medicine, in particular the use of X-rays in medical and dental radio-diagnosis.

In industry, there are a variety of uses for ionising radiation: radiography, detection, measurement, tracing...

A complete inventory of sources of ionising radiation should also include the natural telluric radioactivity (emitted by the subsoil) and cosmic radiation.

6.1. Natural sources of ionising radiation

There are two types of naturally occurring ionising radiation: radiation from space (called *cosmic radiation*) and radiation emitted by the earth's crust, called *telluric radiation*.

6.1.1. Cosmic radiation

Cosmic radiation originates in space (for example from the sun) and strikes the earth's atmosphere. It is comprised of many types of (charged) particles and electromagnetic radiation of highly variable energy, such as protons, neutrons, electrons and photons. Furthermore, the interaction between this cosmic radiation and the constituents of the earth's atmosphere creates radioactive nuclei such as tritium (hydrogen-3) and carbon-14.

Exposure to cosmic radiation increases with altitude (Figure 6.1). Thus flight personnel are the most exposed, particularly those assigned to long-haul flights flying at very high altitudes (up to 11 000 meters). Each flight undergoes a preliminary dosimetric evaluation, taking into account the route to be traveled. For example, a transatlantic flight from Europe to North America imparts an effective dose of 40 µSv (Sievert data: http://www.sievert-system.org/). Astronauts are also subjected to high levels of cosmic exposure. The intensity of comsic radiation also changes with latitude due to the interaction of the charged particles with the earth's magnetic field.

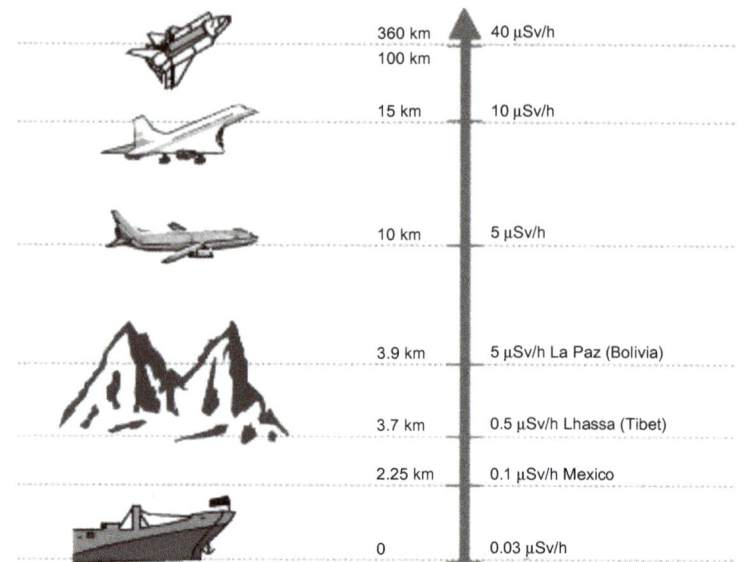

Figure 6.1. Variation of exposure to cosmic radiation as a function of altitude (data: UNSCEAR and IRSN) (drawing: Marion Solvit).

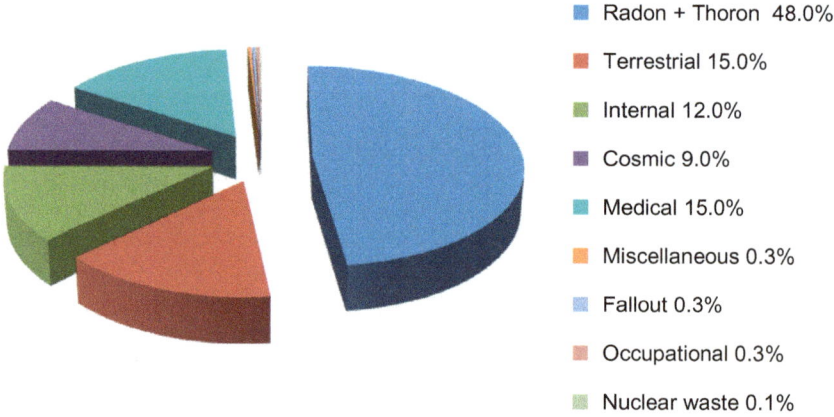

Figure 6.2. Average radiation exposure of the European countries-population (natural and artificial).

6.1.2. Telluric radiation

This radiation comes either from the natural decay chains of uranium-238, uranium-235 and thorium-232, isotopes that are present in terrestrial rocks, or from potassium-40. These isotopes and the corresponding decay products emit both alpha and beta particles. Exposure to the general public originates from external radiation, inhalation (of gasous decay products) and ingestion (from nutrition and drinking water).

In terms of public health, the main component to consider is radon-222, a radioactive gas which is a decay product of uranium-238 and whose own decay products, alpha emitters which settle in the lungs, are carcinogenic.

Exposure to telluric radiation can be considered occupational exposure for certain categories of workers such as miners and hot springs employees. Figure 6.2 shows the average distribution of sources of exposure to ionising radiation in Europe.

It is also possible to measure the average activity due to natural, cosmic or telluric radioactivity for different samples. These values are given in Table 6.1.

Table 6.1. Average activities of different samples due to natural radioactivity (excerpts from "La radioactivité naturelle en 10 épisodes", SFRP Editions, march 1998).

Sample	Average activity	Predominant radioisotopes
Rainwater	0.3 to 1 Bq/L	^{238}U and ^{226}Ra
Sea water	14 Bq/L	90% caused by ^{40}K
Mineral water	2 to 4 Bq/L	dissolved ^{238}U, ^{226}Ra and ^{222}Rn
Milk	80 Bq/L	90% caused by ^{40}K
Potato	150 Bq/kg	90% caused by ^{40}K
Sedimentary soil	1000 Bq/kg	^{238}U, ^{232}Th, ^{40}K
Granitic soil	3000 Bq/kg	^{238}U, ^{232}Th, ^{40}K
Human being (70 kg)	8600 Bq	50% caused by ^{40}K

Further information: radon in the environment and in housing

Radon-222 is a daughter product of the natural decay chain of uranium-238. This chain is summarized in Figure 6.3.

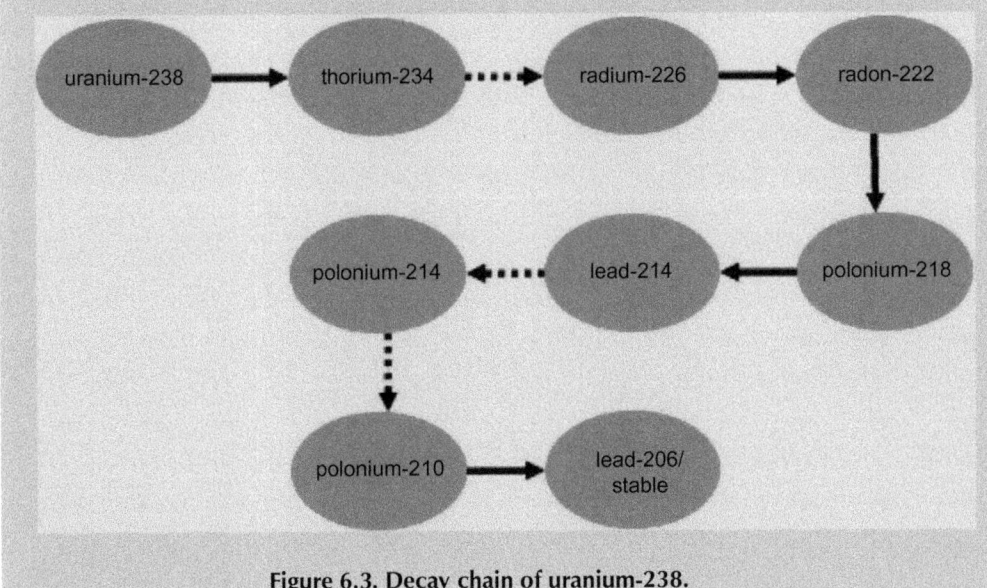

Figure 6.3. Decay chain of uranium-238.

Accordingly, radon-222 is a naturally occurring gas and some of its decay products are alpha particle emitting solids. If inhaled, the lungs are exposed to these alpha particles, which have low penetrating ability but will irradiate the more sensitive cells of the bronchi. Studies conducted on uranium miners have shown that radon exposure significantly increases the risk of lung cancer. Studies have also shown that exposure to radon combined with smoking increases the effects of lung cancer even further.

Therefore, maximum radon concentration levels are defined by the ICRP and European Commision, defining a threshold beyond which it becomes imperative for some buildings to implement radon mitigating solutions.

The radon concentration in the environment, and consequently in houses, varies from one place to another depending on the nature of the subsoil. Granitic soils being richer in uranium than sedimentary soils, the highest radon concentrations are found in those types of geologic environments. Additionally, seasonal and daily variations in the radon concentration can be observed in a single location. Therefore only measurments over longer periods of time should be used when interpreting radon exposure in houses and buildings.

In houses, radon accumulates in enclosed spaces: basements, crawlspaces, and poorly ventilated rooms. Figure 6.4 illustrates how radon infiltrates. It is usually simple and inexpensive to mitigate radon in a home by ventilating the rooms and leak-proofing the basement (by sealing cracks and filling the gaps around pipes).

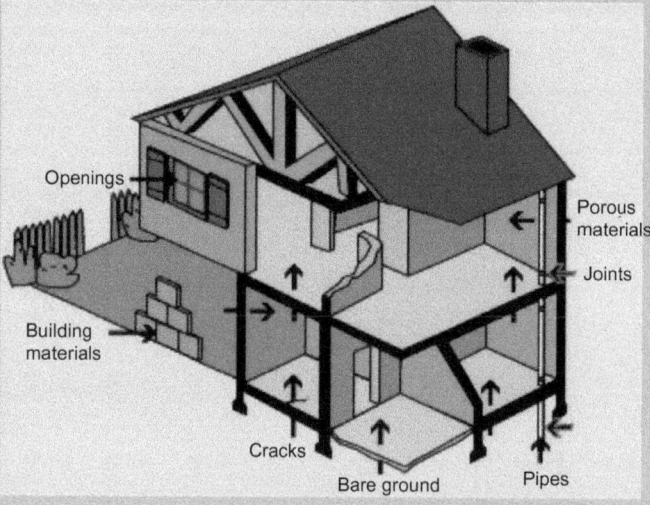

Figure 6.4. Radon infiltration channels in a house (drawing: Marion Solvit).

6.2. Medical applications of ionising radiation

Sources of ionising radiation have two types of applications in the medical field: diagnosis and therapy.

6.2.1. Diagnosis

Here, the practitioner seeks anatomic or metabolic information about a tissue or an organ, and ionising radiation enables him or her to obtain an image or function of this tissue or organ. Two imaging methods using ionising radiation are available, each having its own specific indications: radiology and scintigraphy.

6.2.1.1. Medical and dental radiology

Medical and dental radiological examinations are performed using electrical X-ray generators. These devices do not contain any radioactive material: X-rays are obtained through X-ray tubes which, in the medical field (excluding mammography) operate at high voltages between 50 kV and 140 kV. Radiology is a technique based on the difference of X-ray absorption due to the inhomogeneities of the structure to be examined. While, in the past, radiological film was used as detector, images are now displayed on a monitor thanks to a digital sensor. The duration of X-ray emission is very short, less than 0.1 second, so as to avoid, as much as possible, movement on the patient's or organ's part; however, the beam's dose rate is very high, on the order of 100 mGy.s^{-1} at the patient's level. Naturally, these parameters need to be taken into account in order to optimise staff radiation protection.

X-ray tubes from the medical field are shown in Figure 6.5.

Figure 6.5. X-ray tubes from the medical field.

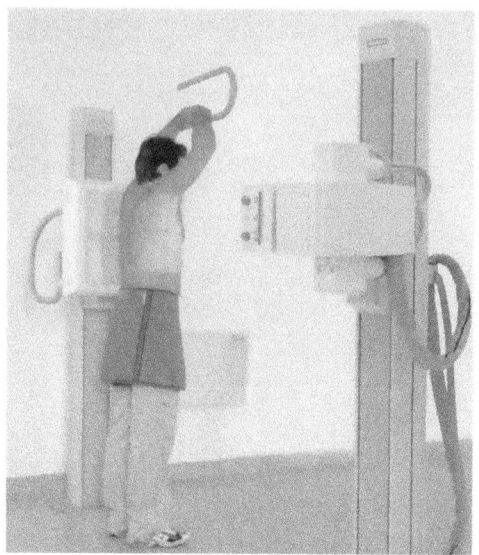
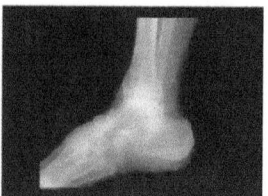

Figure 6.6. X-ray unit (photo: Siemens Healthcare) and X-ray image.

An X-ray unit and an X-ray image are shown in Figure 6.6.

X-ray computed tomography scanners (also called "CT scanners") and mammography devices operate on the same physical principle.

In the case of the CT scanner, the X-ray tube is rotated around the patient in order to obtain cross-section images ("slices") of him/her, which are reconstructed to three-dimensional images (Figure 6.7). The high voltage applied to the tube is on the order of 120 kV.

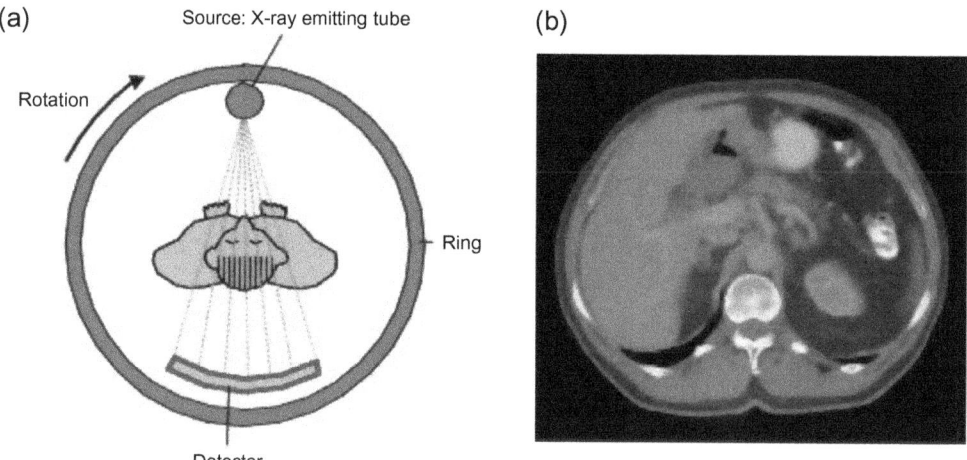

Figure 6.7. Principles of computed tomography (drawing: Marion Solvit) (a) and cross-section obtained from an abdominal examination (b).

In mammography, the structures to be examined are neither very thick nor dense: X-rays must be much less penetrating. The high voltage is therefore lower (about 28 kV) and the anode, which is typically made of tungsten for radiology tubes, is in this case most often made of molybdenum. Figure 6.8 shows a mammography unit and a mammogram.

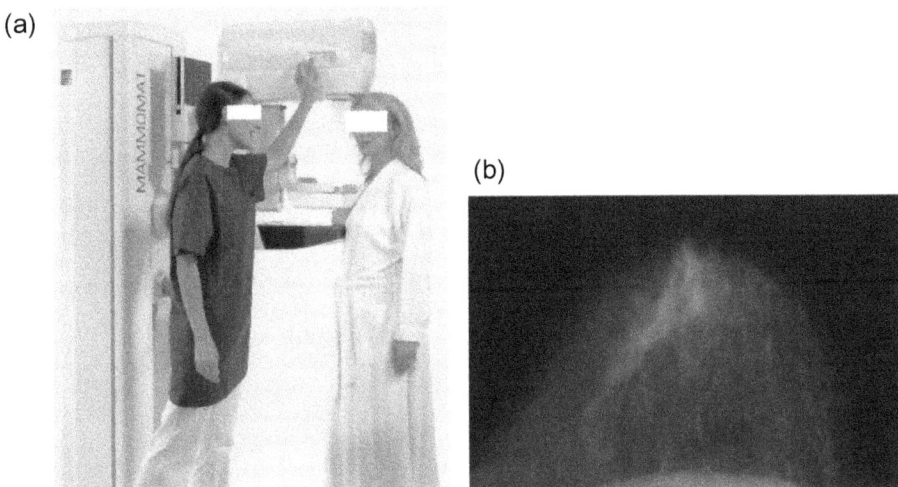

Figure 6.8. Mammography unit (Sources: Siemens Healthcare) (a) and mammogram (b).

In dental radiology, two types of devices are mainly used: generators of the "intraoral" type, intended to take localized snapshots of the teeth, and generators of the "panoramic" type, used to display both entire jaws in a single image. In the first case, the device is a simple X-ray tube (high voltage nearing 70 kV, short emission time) mounted on an articulated arm near the dental operation chair; in the second case, the tube is rotated around the patient's head, and the complete image is obtained in about ten seconds.

Interventional radiology is becoming more generalized: it enables image guiding for delicate surgical procedures. The devices operate on the same principle as those used for diagnosis, but in this case the operator can view the examined area on a screen, "streaming" in real time. In this case, radiation protection of the staff (surgeons, cardiologists, anesthesiologists-radiologists and nurses) is much more difficult to optimise because all personnel are necessarily close to the direct X-ray beam. Therefore adequate training is essential so that, even in the situation of medical emergency they are most often faced with, radiation protection is integrated into their daily practice.

Doses received by radiology operators depend on the unit's manner of use. A study conducted in the veterinarian environment showed that during catheter placement on animals, exposure to the operators' hands (close to the direct beam) can reach dose rates on the order of 50 to 100 $mSv.h^{-1}$. These results can be extrapolated to the medical field, in particular for practitioners performing interventional radiology.

6.2.1.2. Scintigraphy

The principle of scintigraphy is entirely different from that of radiology as described above. Scintigraphy is a "functional" examination, focusing on organ function versus organ morphology. In this case, medication marked with radioactive atoms (called "radiopharmaceutical") is injected in the patient's body, generally intravenously. The radioactive sources used are not sealed: they are liquid (injected, ingested), solid (ingested) or gaseous (inhaled).

The isotopes used are mainly photon emitters: the photons will be able to exit the patient's body, interact with the scintillation detector within a device named "gamma camera". These detectors can be rotated around the patient, to reconstruct a 3D image: SPECT (single photon emission computed tomography) (Figures 6.9 and 6.10). The isotopes used are

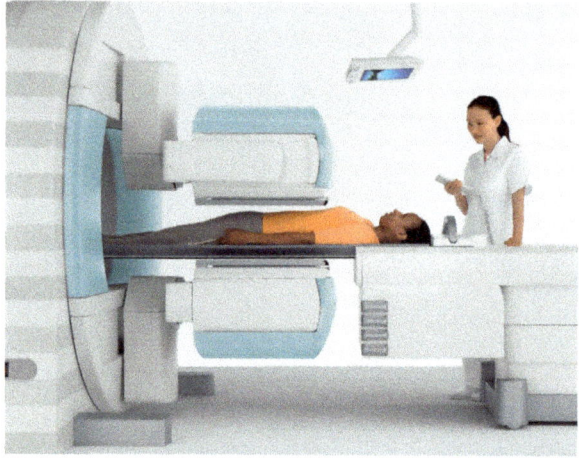

Figure 6.9. Conventional scintigraphy (photo: Siemens Healthcare).

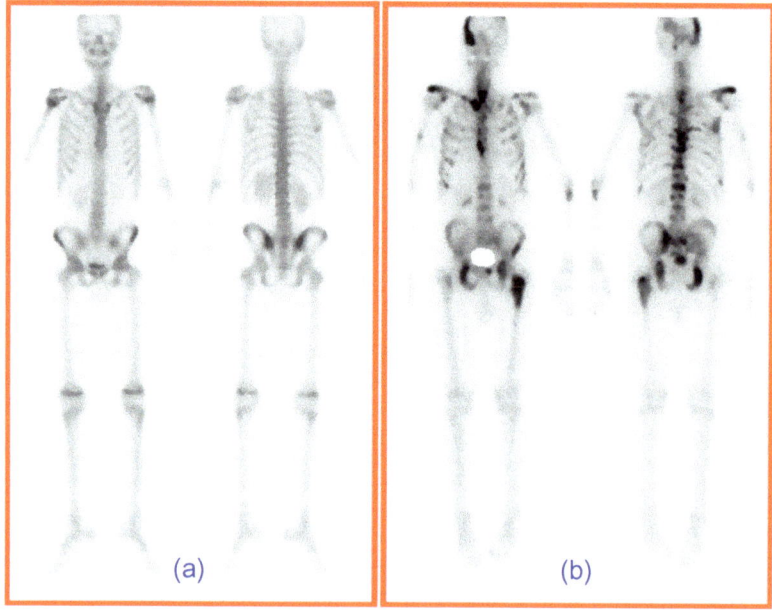

Figure 6.10. Bone scintigraphy using simple photon radio-emitters. Normal image (a). Bone metastases dissemination (b). Credits: CHRU Tours (France).

mainly metastable technetium-99 (90% of all uses), thallium-201, indium-111 or iodine-123. These isotopes all have short half-lives (a few hours to a few days) so as to limit the patient's (and his/her entourage's) exposure. The injected activity depends on the type of examination and on the age and weight of the patient.

Positron emission tomography scanners, often called PET scanners (Figures 6.11 and 6.12) are increasingly common in hospitals and clinics. PET scans require the injection of radiopharmaceuticals marked with a beta+ emitting atom. Today, the most commonly

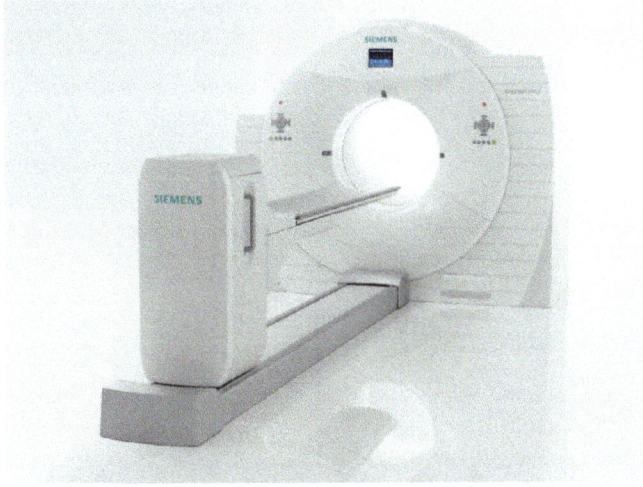

Figure 6.11. Positron emission tomography scanner (photo: Siemens Healthcare).

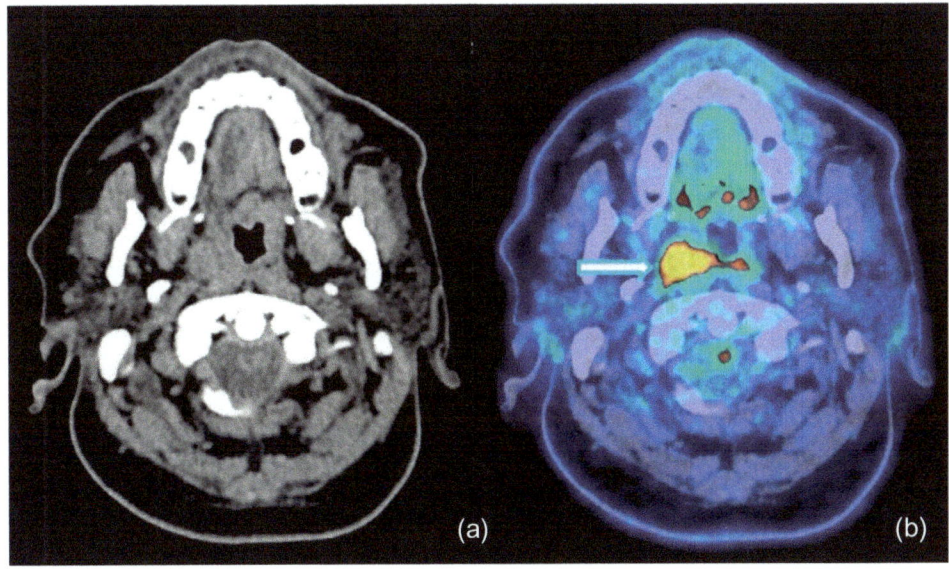

Figure 6.12. Axial PET ^{18}F-FDG image of an oropharynx tumor (white arrow): CT (a) and PET/CT (b). Credits: CHRU Tours (France).

used is fluorine-18. In this case, the two annihilation photons, emitted at a 180° angle, are detected by the detector ring (see Chapter 2, "Interaction of radiation with matter").

In order to superimpose a morphological image and a functional image, the PET scanner and SPECT device is often combined with an X-ray scanner.

In nuclear medicine departments, where these examinations are performed, other procedures also require the use of unsealed radioactive sources:

- metabolic therapy: high activities are administered to the patients with a curative versus diagnostic intent. Here, the most widely used isotope is iodine-131 (administered activity on the order of 4 GBq);

- in vitro diagnosis: marking substances with iodine-125 enables radioimmunoassays.

6.2.2. Therapy

In therapy, the practitioner uses ionising radiation to kill diseased cells, usually cancerous cells. Two techniques exist today: external radiotherapy and brachytherapy.

6.2.2.1. External radiotherapy

Tumors are irradiated by external X-ray beams or high-energy electron beams supplied by an linear accelerator. The choice of beam type and beam ballistics must allow the delivery of a very high dose to diseased tissues while sparing the surrounding healthy tissues as much as possible.

An electron linear accelerator can produce electrons (Figure 6.13) of several MeV and, through bremsstrahlung in a metallic target, X photons of high energy (see Chapter 2, "Interaction of radiation with matter"). Today, they increasingly replace telecobalt therapy devices, which contain a cobalt-60 sealed source of several hundreds of TBq.

6.2.2.2. Brachytherapy

Here, sealed radioactive sources are inserted, as threads or pellets, into the patient in contact with the tissue to be treated. They can be placed directly inside the organ, for example in the prostate, or in a natural cavity such as the uterus or bronchii. The most commonly used isotopes are iridium-192, cesium-137 and iodine-125.

6.2.3. Other equipment

At the edge of the medical field, ionising radiation is also used in the veterinary field (radiology facilities and a few nuclear medicine departments) and to irradiate blood products. This latter technique is designed to eliminate the risk of post-transfusion disease. Irradiators are compact, self-shielded, and contain an irradiation chamber and sealed sources (mainly cesium-137) whose activity ranges between 60 TBq and 180 TBq.

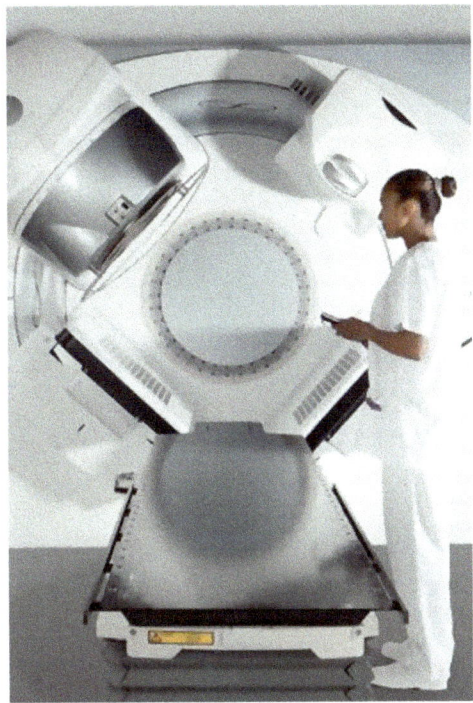

Figure 6.13. Linear accelerator (photo: Elekta-Infinity).

6.3. Industrial applications of ionising radiation

The most common industrial applications of ionising radiation are radiology, metrology (for measurement or calibration purposes) and irradiation. For these applications, sealed radioactive sources, X-ray generators and particle accelerators are used. Unsealed radioactive sources are also useful as tracers in industry, research or environment.

6.3.1. Industrial radiography

Radiography is one of the "non destructive testing" techniques used in industry. Ionising radiation is used in industry to obtain images of metal parts, welds or concrete structures, for inspection purposes and without damaging the structure under examination. It can be performed using X-ray generators or gammagraphy devices.

6.3.1.1. X-ray generators

They operate on the same physical principle as those from the medical field (see Chapter 2, "Interaction of radiation with matter"). They may differ in the tube shape and in the value of the high voltage applied. In industry, some generators operate at high voltages of approximately 400 kV. They can be fixed or mobile.

6.3.1.2. Gammagraphy devices

These units are direct competitors with X-ray generators in the field of non-destructive testing. They can also be fixed or mobile. They contain a sealed radioactive source, a source enclosed in a casing designed to avoid any material dispersion. This source is ejected from its container into a sheath in order to perform the gammagraphy (Figure 6.14).

(a)

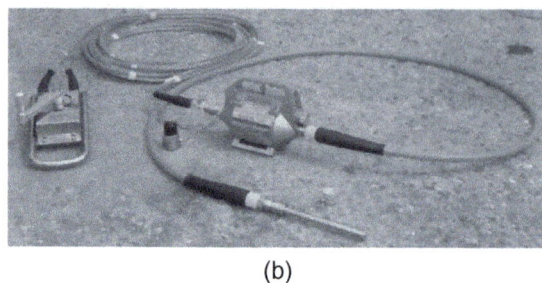
(b)

Figure 6.14. Gammagraphy device only (a) and with its accessories (b) (photos: Cegelec).

Figure 6.15. RAPISCAN 627XR baggage screening unit (photo: HTDS).

The most common sources are iridium-192 (3 to 4.5 TBq) or cobalt-60 (1 to 2 TBq) which is used when the thickness to be penetrated is more important, for example when dealing with concrete (bridges, buildings).

6.3.1.3. Baggage screeners

Luggage, personal carry-on items and cargo are controlled in airports, train stations and other security-sensitive facilities. The VHV varies from 50 to 300 kV. X-ray generators are contained in self-shielded tunnels, and the device includes a conveyor belt on which the objects to be inspected are placed. The dose rate at the tunnels' entry and exit points is less than 1 $\mu Sv.h^{-1}$.

Lorries and ship containers are screened with high power generators (up to several megavolts). The absorbed dose for a complete scan is on the order of 50 to 80 μSv. Figures 6.16 and 6.17 quad show the installation and corresponding results.

Figure 6.16. Vehicle screening equipment: "Eagle Mobile" Rapiscan Cargo scanner (photo: HTDS-RAPISCAN).

Figure 6.17. Section of a lorry and content detection (photo: HTDS-RAPISCAN).

6.3.2. Metrology and analysis devices

Numerous devices contain sources of ionising radiation (radioactive source or X-ray generator) for metrological or analysis purposes.

6.3.2.1. Cristallography generators

Diffractometry relies on the proximity of the wavelength of X-rays (produced by an electric X-ray generator operating at low values of high voltage) to interatomic distances. A "direct" X-ray beam irradiates the sample. The scattering angles and the intensity of the scattered radiation are then analysed (figure 6.18). These generators are used for analysis in the fields of metallurgy, geology, etc. in order to determine the structure of the crystals or molecules in a sample.

6.3.2.2. X-ray fluorescence alloy analysers

X-ray fluorescence analysers are widely used to quickly determine the composition of alloys or paint (see: lead paint detectors). Given the high sensitivity and high precision of these devices, they are also used in laboratories to test samples, in particular to check for impurities. The most commonly used radionuclides are iron-55, cadmium-109 and cobalt-57.

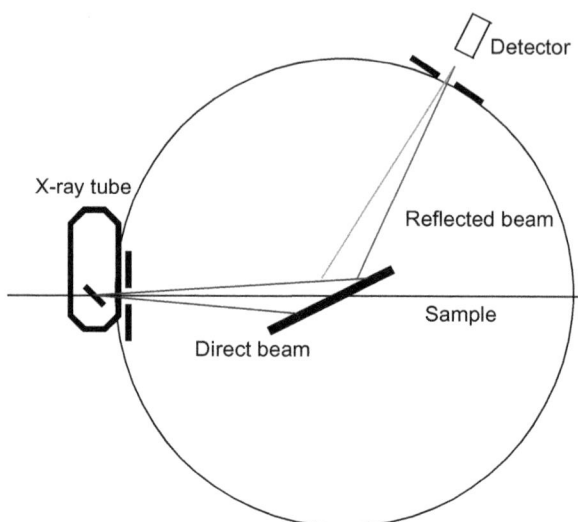

Figure 6.18. Operating principles of a diffractometer.

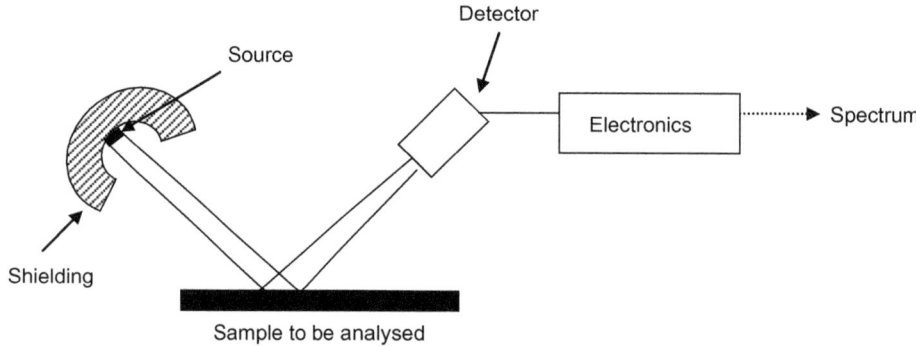

Figure 6.19. Operating principles of an X-ray fluorescence analyser.

This analytical technique is based on the excitation of the atoms in the analysed sample and the analysis of their characteristic X-ray lines. This X-ray fluorescence results from the photoelectric effect on the target atoms and it is therefore necessary that the sources used emit low-energy X or γ radiation (iron-55: 6 keV, cadmium-109: 22 keV, cobalt-57: 15 keV). The energies of the detected X-ray lines thus indicate which elements are present in the analysed sample, while the peak heights show the amount present (Figure 6.19).

This technique, which allows both qualitative and quantitative measures, is especially used in the chemical industry and in metallurgy to measure tinning or galvanizing, to analyse alloys and to sort scrap metal.

Further information: lead paint detectors

In order to prevent lead poisoning when using paint, specific detectors are used to detect the presence of this heavy metal. The measurements must be carried out in buildings when a case of lead poisoning is reported or prior to the sale of an old property, located in a risk area. They must be performed with a portable X-ray fluorescence device such as the one presented in Figure 6.20.

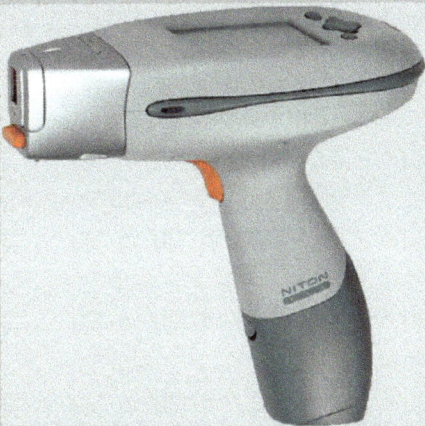

Figure 6.20. Example of lead paint detector (Photo: Fondis).

Such detectors operate on the same principles as the X-ray fluorescence analysers described above. They contain cadmium-109 or cobalt-57 sources, that have an activity of the order of 400 MBq. They enable the measurement of low levels of lead, below the legal threshold set at 1 mg.cm^{-2}.

These measurements can be performed by a wide variety of professionals: control agencies, architects, notaries and real estate agents.

6.3.2.3. Electron capture detectors

Electron capture detectors are used widely in gas chromatography to determine the impurity levels of solutions. The maximum activity of the β radiation emitting sources used is of the order of 500 MBq for nickel-63 and 7.4 TBq for tritium.

The gas coming from the chromatograph passes through an ionisation chamber containing a source of nickel-63 (or tritium), beta radiation emitter which ionises the medium. When a component containing impurities passes through the chamber, it combines with the free electrons present and the ionisation current drops. The concentration of the analysed component will be proportional to the decrease in ionisation current (Figure 6.21).

6.3.2.4. Thickness gauges

Thickness measurements can be performed with gauges containing sealed radioactive sources. Depending on the applications, two different techniques can be used:

- β or γ transmission,

- β or γ backscatter.

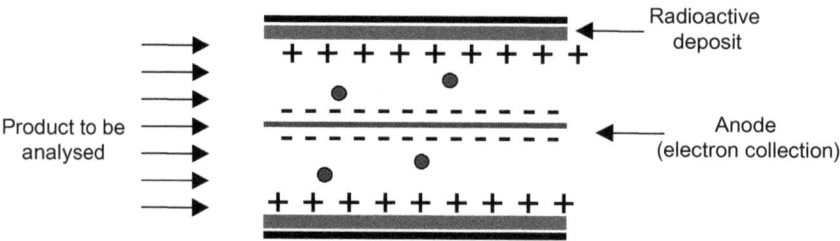

Figure 6.21. Operating principles of an electron capture detector.

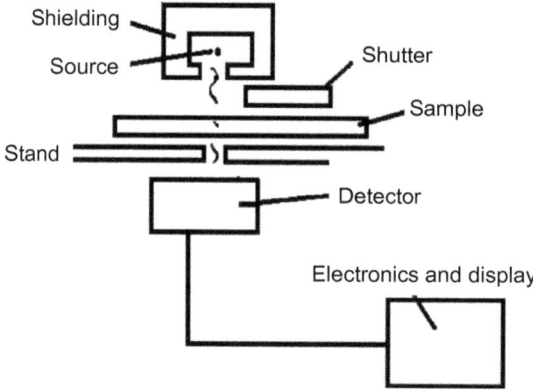

Figure 6.22. Operating principles of a transmission thickness gauge.

In the first case, source and detector are situated on either side of the specimen to be tested (Figure 6.22). The detector measures the radiation transmitted behind the sample: its intensity is proportional to the thickness and to the atomic number of the material.

In the case of backscattering gauges, source and detector are situated on the same side of the sample to be analysed (Figure 6.23). The highly collimated beam irradiates the specimen. The measured intensity of the backscattered radiation is again proportional to the thickness and the atomic number of the material.

Transmission measurements are most commonly used. The backscattering technique is limited to cases where access to opposite sides of the sample is difficult (for example, to measure the thickness of bitumen coating) or to measure the thickness of deposits (jewellery, electronics).

The most frequently used radionuclides are:

- β transmission or β backscattering: strontium-90+yttrium-90, krypton-85, prometheum-147 and thallium-204 (measurements of thin paper and metal rubber sheets, substrate coatings, gold deposits on connectors, electronics and jewellery);

- γ transmission or backscattering: americium-241 and cesium-137 (metal sheets, pipe and tank walls, glass, for thicknesses greater than 1 g.cm^{-2}).

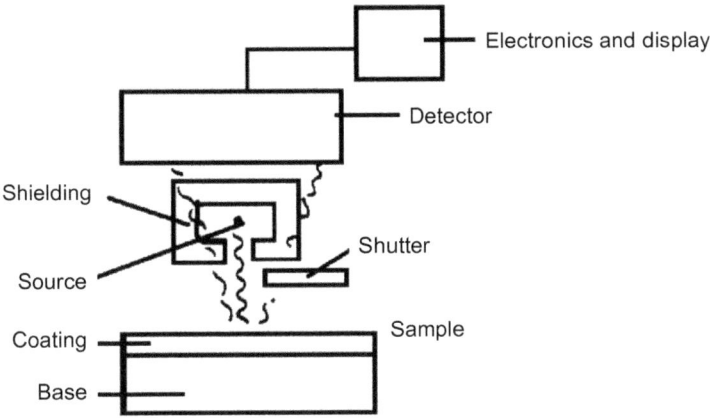

Figure 6.23. Operating principles of a backscattering thickness gauge.

6.3.2.5. Fill level gauge

A sealed gamma emitting radioactive source and a detector are placed on either side of the container whose fill level is to be gauged (Figure 6.24).

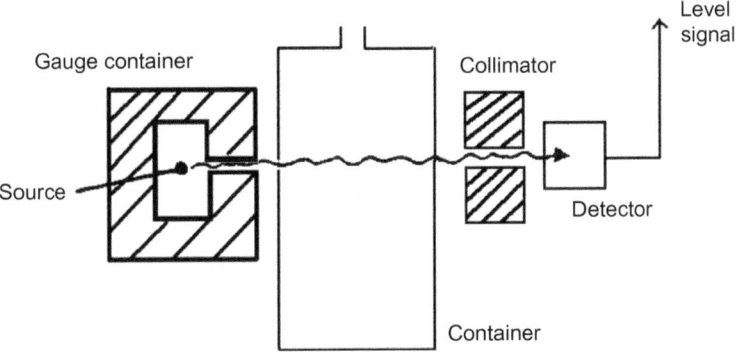

Figure 6.24. Operating principles of a level gauge.

The intensity of the radiation sensed by the detector decreases when the fill level reaches the horizontal line marking the path of the beam. This variation in intensity can trigger specific actions such as activating an alarm, stopping the filling or initiating a manufacturing process. Level measurements involving radioactive sources are used to implement a wide variety of processes, when:

- highly accurate measurements of the fill level are required: beverage and perfume bottling;
- containers are opaque: chemical industry, filling gas tanks, filling metallic cans with drinks.

The most commonly used radionuclides for these applications are cobalt-60 and cesium-137 for dense and thick materials, and americium-241 for bottle and can liquids.

6.3.2.6. Measures of soil humidity and density

Neutron emitting sources are used to determine the nature of the subsoil in exploration activities (water or hydrocarbon exploration). Neutrons are slowed down by collisions with light atoms (hydrogen in particular) that are present in the medium. The richer in hydrogen the medium is, the more backscattered the neutrons will be, making them easy to detect by a nearby neutron counter.

Most often, this involves neutron emitting sources of americium-241 and beryllium-9 of a few GBq.

Cesium-137 sources are used in civil engineering to determine the soil density or the level of compaction on road and rail construction sites. Detection of the gamma radiation emitted by cesium-137 enables determination of the density of the soil or rocks by calculating the attenuation of the radiation caused by the absorbing medium. The source activity ranges from 1 GBq to a few GBq.

6.3.3. Industrial irradiators

Industrial irradiation techniques involve exposing samples to radiation (usually photons or electrons) in order to preserve or improve some of their properties. For example, in the chemical industry, these techniques are used for plastics crosslinking and hardening. In the bio-medical field, they are used to sterilise medical and surgical products. Additionally, food products can be sterilised in the same way in order to increase their shelf life.

The devices used are linear particle accelerators or irradiators containing high activity radioactive sources of cobalt-60 or cesium-137. Figure 6.25 shows a diagram of an electron accelerator based ionisation installation.

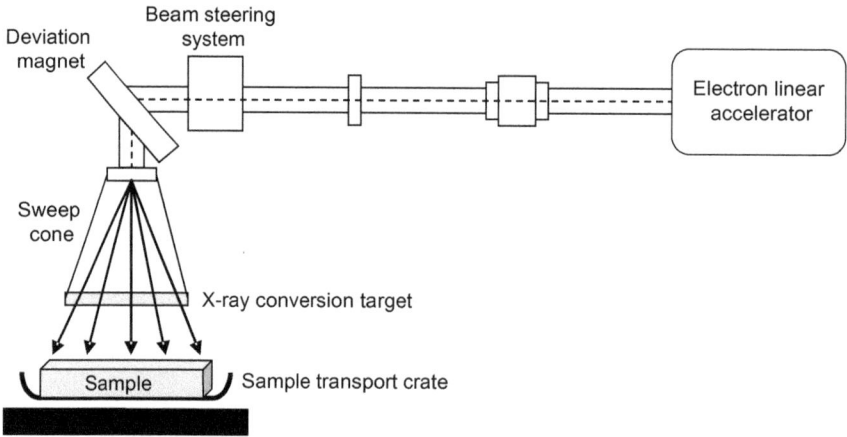

Figure 6.25. Diagram of an electron beam ionisation installation.

6.3.4. Miscellaneous uses of radionuclides as sealed sources

Monitoring services such as radioprotection and dosimetry services frequently use sealed sources of variable size and shape, of low activity (a few kBq to a few MBq), intended to calibrate ionisation radiation detectors (radiation protection devices, for example).

Some smoke detectors are equipped with an americium-241 source of about 37 kBq. They are safe under normal operating conditions (if they are not disassembled). However, in order to keep these devices under control, their sale to the general public is prohibited and they may only be installed in buildings' common areas. Regulators are considering a gradual discontinuation of these detectors starting in 2007.

In conclusion of this "Miscellaneous" section, lightning rods equipped with radium-226 or americium-241 radioactive sources must be mentioned. They are, however, becoming obsolete.

6.3.5. Uses of radionuclides as unsealed sources in industry and research

The use of unsealed sources outside the medical field is extremely diverse, with applications in research, industry or earth sciences.

Used as tracers, they enable hydrology studies, process control in the chemical industry, monitoring of the wear on mechanical parts, and the search for pipe leaks. In biological research, these sources are used as markers in the study of biological processes.

6.3.5.1. Applications of unsealed sources in industry

Radioactive tracers in liquid or gaseous form are used to study mass transfers in natural and industrial environments. In all cases, the goal is to take spatial and quantitative measurements of the radioactivity which has dispersed in the environment under study. There are various applications of this technique. Some examples include:

- subterranean hydrology: determining the speed and direction of water flow;

- surface hydrology: monitoring the dispersion of pollutants, searching for leaks in dams and channels;

- dynamic sedimentology: measuring sediment displacement, monitoring seabed transport, optimising the outfall point of dredged materials;

- industrial tracing: searching for leaks in sealed objects, assessing wear on equipment subject to intense friction, evaluating oil consumption.

Several radioactive isotopes can be used, such as carbon-14, chlorine-36, nitrogen-15, oxygen-18, potassium-40 and all of the isotopes from natural radioactive chains. Tritium remains the most widely used isotope: as tritiated water, it is used to study hydrodynamic flows (subsoil water, lakes).

6.3.5.2. Applications of unsealed sources in biological and medical research

In this area, the tracers most often used are pure beta emitters, although some X or gamma emitters are also employed. Applications include:

- marking of DNA (phosphorus-32, phosphorus-33, sulphur-35);

- marking of iodized proteins such as thyroid hormones (iodine-125 and iodine-131);

- enzymology (calcium-45);

- developing radiopharmaceuticals (technetium-99m and all isotopes used in nuclear medicine);

- marking of red blood cells (chromium-51).

Carbon-14 and potassium-40 are also used to date objects and biological samples.

6.4. Civil nuclear industry

All industries involved in the entire nuclear fuel cycle, from uranium extraction and enrichment, nuclear power plant operations, to nuclear power plant waste processing, constitutes the "civil nuclear industry" (according to the CEA "Dossiers thématiques" n° 6 and n° 7).

6.4.1. Nuclear fuel

Fuel is a material which provides heat by burning. The best known fuels are wood, coal, natural gas and petroleum. By analogy, the uranium used in nuclear power plants is called "nuclear fuel", as it also releases heat, but in this case by fission instead of combustion (see Chapter 2, "Interaction of radiation with matter").

After use in the reactor, nuclear fuel can be processed to extract recyclable energetic material, thus the term "nuclear fuel cycle". This cycle is summarized in Figure 6.26 and includes the following industrial operations:

- uranium mining,

- fuel fabrication,

- use of fuel in the reactor,

- reprocessing spent fuel discharged from the reactor,

- waste processing and storage.

When comparing the same amounts, nuclear fuel provides much more energy than fossil fuel (coal or petroleum). Used in a pressurised water reactor (or PWR, see 6.4.4), one kilogram of uranium produces 10 000 times more energy than one kilogram of coal or petroleum in a thermal power plant. Moreover, nuclear fuel remains viable in the reactor for a long time (several years), unlike conventional fuels which are quickly exhausted. Nuclear

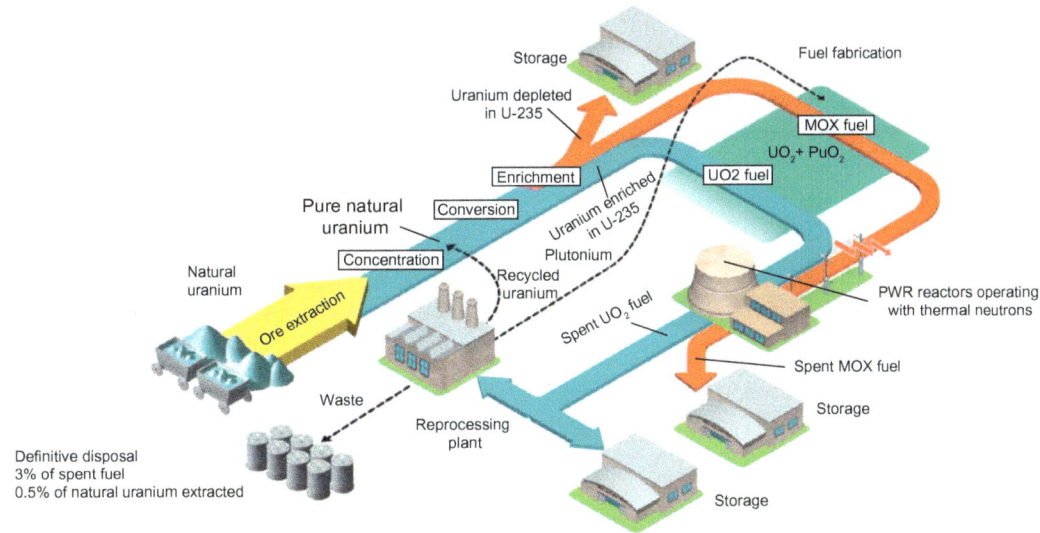

Figure 6.26. Simplified diagram of the fuel cycle in France today.

fuel further differs from other types of fuel in that it must undergo numerous processing operations after extraction before use in the reactor.

For simplicity's sake, we will limit the following discussion to nuclear fuel used in PWRs, the most common type of reactor in Europe.

6.4.2. Uranium ore extraction

Uranium is a relatively common metal in the earth's crust (50 times more common than mercury, for example). It is found in rocks known as "uraniferous", such as uraninite and pitchblende. The nuclear fuel cycle thus begins with the mining of uranium ore in open pit or underground mines. The known ore deposits are found primarily in Australia, the United States, Canada, South Africa and Russia.

6.4.3. Nuclear fuel fabrication

6.4.3.1. Uranium concentration and refining

The uranium content of natural ore is generally quite low, so it is essential to concentrate it. Rocks are first crushed and finely milled, before uranium is extracted by various chemical operations. The produced concentrate has the appearance of a yellow paste and is called "yellow cake". It contains about 75% uranium. This uranium concentrate cannot be used as is in nuclear reactors. It must first be purified through various refining stages. When it reaches optimum purity, it is converted to uranium tetrafluoride (UF_4), then uranium hexafluoride (UF_6).

6.4.3.2. Uranium enrichment

Fueling PWRs requires fuel with a uranium-235 content between 3% and 5%. Only this uranium isotope can undergo nuclear fission, thereby relasing energy (see Chapter 2, "Interaction of radiation with matter"). However, in natural uranium, the relative proportions are 99.3% of uranium-238 and 0.7% of fissile uranium-235. The process by which the proportion of uranium-235 is increased is called "enrichment": it enables the concentration of uranium-235 to increase from 0.7% to approximately 3%.

6.4.3.3. Fuel assemblies

After enrichment, uranium hexafluoride is converted to uranium oxide in the form of a black powder. This powder is then compressed and sintered (baked) in order to produce small cylinders with a height of approximately 1 cm, called "pellets". Each pellet, weighing only 7 g, is capable of releasing as much energy as one ton of coal. The pellets are inserted in 4 meter long metallic tubes made of a zirconium alloy in order to form the fuel rods (Figure 6.27).

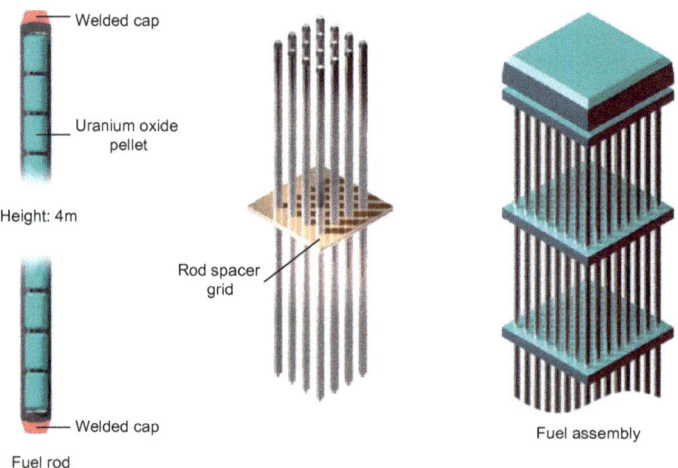

Figure 6.27. Preparing fuel assemblies.

For a single power plant, over 40 000 rods are prepared and collected into "bundles" of square section called fuel assemblies. Each assembly contains 264 rods. Loading a 900 megawatt nuclear reactor requires 157 assemblies containing a total of 11 million pellets.

Fuel assemblies are arranged in a precise configuration to form the core of the nuclear reactor. Each assembly will remain there for three to four years. During this period, uranium-235 fission will provide the heat necessary to produce steam and electricity (see 6.4.4).

6.4.4. "Pressurised Water Reactor" type nuclear reactor

6.4.4.1. Principles of operation

A power plant is a facility which produces electricity. There are thermal power plants, hydro-electric power plants... and nuclear power plants. They all operate on the same principle: a rotating turbine is coupled to an alternator that produces electricity. In nuclear power plants, uranium fuel replaces the fossil fuel (petroleum, coal or gas) used in thermal power plants.

The principle on which such power plants rely is nuclear fission – here, the fission of uranium-235 atoms – (see Chapter 2, "Interaction of radiation with matter"). The fission products carry a large part of the energy released by the reaction. This energy is dissipated very quickly as heat due to the impact of the fission products with the surrounding matter. Their initial energy is eventually converted into heat: locally, the uranium temperature increases. The principle of a nuclear power plant is to recover this heat in order to transform it into electricity. Each fission produces, on average, two to three neutrons which will, in turn, be able to trigger new fissions and the release of new neutrons, and so on... it is a chain reaction (see schematic diagram in Figure 6.28).

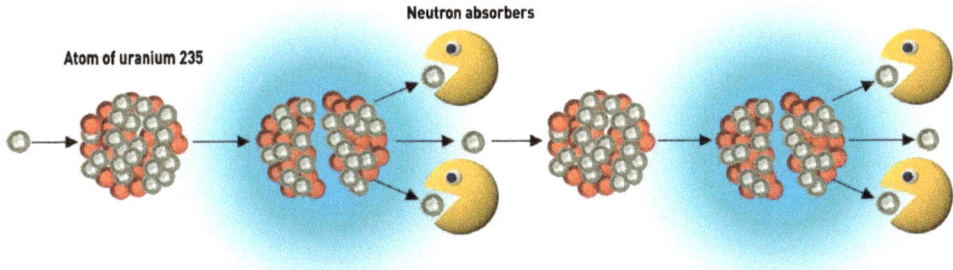

Figure 6.28. Controlled chain reaction in a nuclear reactor.

In a nuclear reactor, the chain reaction is controlled in order to maintain a constant rate of fissions: of the two to three neutrons released per fission, only one triggers a new fission while the other two are captured. A balance must be reached: one fission triggers one fission, which triggers one fission, which triggers one fission, and so on. The amount of heat released at each second in the uranium mass is thus perfectly controlled.

The nuclear fuel is placed in the nuclear reactor core (see 6.4.3.3) shown as a diagram in Figure 6.29. Uninterrupted control of the chain reaction is ensured by neutron absorbing control rods made of, for example, boron. These rods are mobile within the reactor core: they can be partially inserted or removed depending on the number of neutrons to be absorbed. They allow control of the reactor. In addition, in case of an accident, inserting the rods completely (i.e. dropping them) into the fuel stops the chain reaction almost instantaneously.

In order to produce electricity, the energy released as heat during the fission of uranium-235 nuclei must be recovered. This is accomplished by the coolant, in the case of a PWR a liquid that transports the heat. By circulating around the uranium rods, the coolant plays two roles: it removes heat from the fuel to transport it out of the reactor core, and maintains the core temperature at a value which preserves the integrity of the materials.

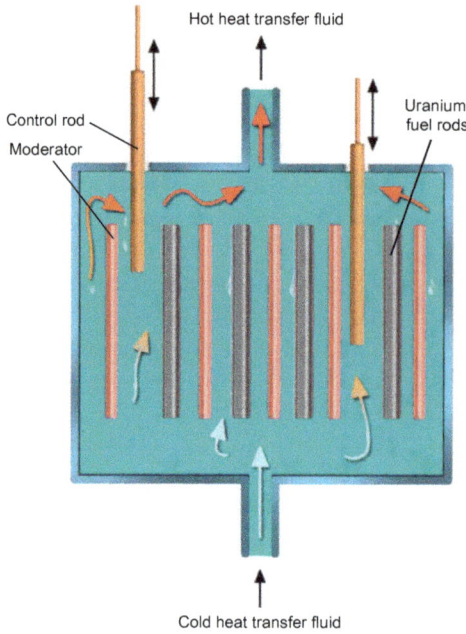

Figure 6.29. Core of a nuclear reactor.

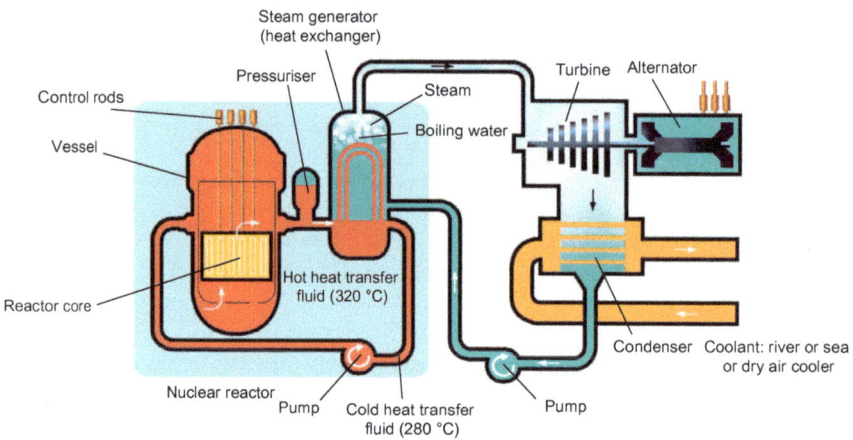

Figure 6.30. Diagram showing the operation of a pressurised water reactor.

Figure 6.30 shows the schematic diagram of a PWR. The nuclear boiler, located in the reactor, is the part of the nuclear power plant which provides the heat necessary for the production of steam. Other components (turbine, alternator, etc.) are common to all power plants.

There are other categories of nuclear reactors, which differ from PWRs according to the types of fuel, coolant and moderator (which slows down the neutrons) they use.

6.4.4.2. Radionuclides present in a reactor

In addition to the uranium forming the fuel (and whose fission properties are used), the radioactive elements found in a nuclear reactor are:

- β and γ emitting fission products (iodines, cesium-137, strontium-90, ...);
- activation products resulting from the absorption of neutrons by the fuel cladding materials, by the coolant's impurities and by various structural elements which become radioactive. They are also β and γ emitters (cobalt-60, for example);
- α emitting transuranic elements which can be found in the fluid circuits in case of severe rupture of the fuel cladding.

These products may circulate in the primary circuit and deposit preferentially in areas where the flow of cooling water is slowed. They can therefore cause significant dose rates and risk of external exposure.

These products can also cause surface and atmospheric contaminations when an intervention is required on open pipes or capacitors. The risk of internal exposure may then occur.

Under normal operation conditions, significant dose rates due to γ and neutron radiation are present in the vicinity of the reactor vessel.

6.4.5. Nuclear fuel reprocessing

Over time, nuclear fuel transforms:

- uranium-235 is progressively consumed due to the fission reactions;
- decay products appear (because they absorb neutrons, these products disrupt the chain reaction).

Therefore, after a certain time (approximately 3 years), the fuel must be removed from the reactor even if it still contains significant amounts of recoverable energy materials, in particular uranium and plutonium. This spent fuel is highly radioactive due to the presence of fission products; it is therefore stored in a cooling pond near the reactor for three years to allow for its activity to decrease before reprocessing.

Reprocessing involves:

- recovering still usable materials, plutonium and uranium, to be used again for electricity production. This constitutes the recycling of energetic materials contained in spent fuel;
- sorting of non-recyclable radioactive waste.

Some countries, for example Sweden and the United States, have chosen not to reprocess. There, spent fuel is considered waste and disposed of directly after removal from the reactor. The countries who have chosen to establish reprocessing plants are France (AREVA NC La Hague plant), the United-Kingdom, Russia and Japan. Other countries such as Germany, Switzerland and Belgium have their spent fuel reprocessed abroad (in particular in France).

Bibliography

Antoni R., Bourgois L., *Physique appliquée à l'exposition externe : dosimétrie et radioprotection,* Springer, 2012.
Attix F.H., *Introduction to Radiological Physics and Radiation Dosimetry,* 2nd ed., Wiley, 1986.
Cember H., Johnson T.E., *Introduction to Health Physics,* 4th ed., Mc Graw Hill Companies, 2009.
Chelet Y., *La radioactivité manuel d'initiation,* Nucléon, 2006.
Council Directive 2013/59/Euratom laying down basic safety standards for protection against the dangers arising from exposure to ionising radiation, 2013.
Delacroix D., Guerre J.-P., Leblanc P., *Radionucléides & Radioprotection : Guide pratique,* EDP sciences, 2006.
Foos J., *Manuel de radioactivité,* Hermann, 2009.
Forshier S., *Essentials of Radiation, Biology and Protection,* 2nd ed., Delmar, 2008.
ICRP Publication 60, *1990 Recommendations of the International Commission on Radiological Protection,* Pergamon, 1991.
ICRP Publication 84, *Pregnancy and Medical Radiation,* Pergamon, 2000.
ICRP Publication 103, *2007 Recommendations of the International Commission on Radiological Protection,* Elsevier, 2007.
ICRP Publication 107, *Nuclear Decay Data for Dosimetric Calculations,* Elsevier, 2009.
Jimonet C., Metivier H., *Personne Compétente en Radioprotection, principes de radioprotection-réglementation,* 2nd ed. EDP sciences, 2009.
Johns H.E., Cunningham J.R., *The Physics of Radiology,* 4th ed., Charles C. Thomas, 1983.
Knoll G.F., Radiation Detection and Measurement, 4th ed., Wiley, 2010.
Magill J., Galy J., *Radioactivity Radionuclides Radiation: with the fold-out Karlsruhe chart of the nuclides,* Springer, 2005.
Martin A., Harbison S., Beach K., Cole P., *An Introduction to Radiation Protection,* 6th ed., Hodder Arnold, 2012.
Martin J.E., *Physics for Radiation Protection: A Handbook,* 2nd ed., Wiley, 2006.
Noz Marilyn E., et al., *Radiation protection in the health sciences,* 2nd ed., World Scientific, 2007.
Seeram E., *Radiation Protection,* Lippincott, 1997.
Shapiro J., *Radiation Protection: A Guide for Scientists, Regulators and Physicians*, Harvard University Press, 2002.
Stabin M.G., *Radiation Protection and Dosimetry: An Introduction to Health Physics,* Springer, 2008.
Statkiewicz Sherer M.A., Visconti P.J., Russell Ritenour E., *Radiation Protection in Medical Radiography,* Elsevier, 2006.
Van den Eijnde J., Schouwenburg M., *Practical Radiation Protection,* Syntax Media, 2013.
Vivier A., Lopez G., *Calculs de doses générées par les rayonnements ionisants principes et utilitaires,* EDP sciences, 2012.

www.ingramcontent.com/pod-product-compliance
Ingram Content Group UK Ltd.
Pitfield, Milton Keynes, MK11 3LW, UK
UKHW062045180426
11947UKWH00030B/2050